A Guide to Bioethics

POCKET GUIDES TO
BIOMEDICAL SCIENCES

https://www.crcpress.com/Pocket-Guides-to-Biomedical-Sciences/book-series/CRCPOCGUITOB

The ***Pocket Guides to Biomedical Sciences*** series is designed to provide a concise, state-of-the-art, and authoritative coverage on topics that are of interest to undergraduate and graduate students of biomedical majors, health professionals with limited time to conduct their own searches, and the general public who are seeking for reliable, trustworthy information in biomedical fields.

A Guide to Bioethics

Emmanuel A. Kornyo

CRC Press is an imprint of the
Taylor & Francis Group, an **informa** business

CRC Press
Taylor & Francis Group
6000 Broken Sound Parkway NW, Suite 300
Boca Raton, FL 33487-2742

Printed on acid-free paper

International Standard Book Number-13: 978-1-138-63198-4 (Paperback)

Library of Congress Cataloging-in-Publication Data

Names: Kornyo, Emmanuel A., author.
Title: A guide to bioethics / Emmanuel A. Kornyo.
Other titles: Pocket guides to biomedical sciences.
Description: Boca Raton : CRC Press/Taylor & Francis, 2018. | Series: Pocket guides to biomedical sciences
Identifiers: LCCN 2017012786| ISBN 9781138631984 (pbk. : alk. paper) | ISBN 9781138632011 (hardback : alk. paper)
Subjects: | MESH: Bioethical Issues | Biotechnology--ethics | Bioengineering--ethics | Ethics, Medical
Classification: LCC QH332 | NLM WB 60 | DDC 174.2--dc23
LC record available at https://lccn.loc.gov/2017012786

Visit the Taylor & Francis Web site at
http://www.taylorandfrancis.com

and the CRC Press Web site at
http://www.crcpress.com

Contents

Series Preface

Dramatic breakthroughs and nonstop discoveries have rendered biomedicine increasingly relevant to everyday life. Keeping pace with all these advances is a daunting task, even for active researchers. There is an obvious demand for succinct reviews and synthetic summaries of biomedical topics for graduate students, undergraduates, faculty, biomedical researchers, medical professionals, science policymakers, and the general public.

Recognizing this pressing need, CRC Press has established the *Pocket Guides to Biomedical Science* series, with the main goal to provide state-of-the-art authoritative reviews of far-ranging subjects in short readable formats intended for a broad audience. Volumes in the series will address and integrate the principles and concepts of the natural sciences and liberal arts, especially those relating to biomedicine and human wellbeing. Future volumes will come from biochemistry, bioethics, cell biology, genetics, immunology, microbiology, molecular biology, neuroscience, oncology, parasitology, pathology, and virology, as well as other related disciplines.

In this volume, Dr. Emmanuel A. Kornyo focuses on bioethics, a contentious topic brought about by continuing advances in biology, medicine, and engineering that appear to stretch the boundary between pertinent medical needs and widely accepted moral values. Given the ostentatious absence of a comprehensive, yet jargon-free book on bioengineering and genetics from a bioethics and philosophical perspective, this volume represents a timely effort to fill the knowledge vacuum. The goal of this volume is the same as the goal for the series—to simplify, summarize, and synthesize a complex topic so that readers can reach to the core of the matter without the necessity to carry out their own time-consuming literature searches.

We welcome suggestions and recommendations from readers and members of the biomedical community for future topics in the series and experts as potential volume authors/editors.

Dongyou Liu, PhD
Sydney, Australia

Preface

A Guide to Bioethics: *Pocket Guides to Biomedical Science*

The field of biotechnology has seen an unprecedented growth and development. A better understanding of molecular biology especially nucleic acids (DNAs and RNAs) and the ability to process high-throughput biodata has galvanized a new wave of scientific vigor and rigor. A voluminous record of scientific publications in molecular biology and biotechnology continues to elucidate the very foundational basis for the emergence, development, and the complexity undergirding life. Indeed, new biological tools and techniques have made it plausible to bioengineer and potentially generate new forms of life while the field of biopharmaceutical has also seized these arrays of opportunities to develop the next generation of biologics to cure very debilitating illnesses hitherto incurable in the annals of healthcare. This new vim has synergistic effects—sometimes vitiating on social conventions, regulatory policies, biomedicine, forensics, diagnostics, and the development of novel products and the creation of wealth among an array of others. Axiomatically, these prospects have also exacerbated some ethical and moral quagmires. The book dexterously examines the biotechnological trajectory through the lenses of bioethics and some perspectives on the emerging and merging issues of scientific advancement. *A Guide to Bioethics* explores the social contexts, the nature and the context of science, the roles of society in shaping the scientific and biotechnological enterprise as well as the therapeutic prospects of personalized medicine. *A Guide to Bioethics* and the subsequent series is a tacit invitation to ponder on the candidness of the seismic breakthroughs in biological sciences, findings regarding the significance and the limits of the scientific process, particularly, biotechnological innovations! As a result, the primary audience: students of bioethics and biotechnology, environmental scientists, pre-med and medical students, physicians, nurses and ancillary medical professionals will find this book useful. I believe the policy aspects of the book will be a good read for seasoned policy makers, IRBs (Institutional Review Boards), public health practitioners especially those concerned with Genome Wide Associative Research (GWAR), clinical researchers (pharmacogenomics in particular) and anyone with a flair for bioethics.

Structure and Scope

The book is structured around four sections in scope. Section I (Chapter 1) examines the significance of biotechnology at various epochs. I present a brief synopsis of the nature of science and its paradigmatic implication for contextualizing the emergence of biotechnology. The scientific and historic analyses lay the fodder for the emergence of bioethical norms in biomedical research. In this chapter based on historic evidence, readers will discover why biotechnology may be construed to be the oldest of the sciences. Very old civilizations and cultures such as the Ancient Near East (ANET), China, Egypt, Songhai Empire, and Aztec had developed elaborate technologies about fermentation used in preserving and making food as well as medicine. It does appear that each generation has advanced and adapted these technologies culturally. Molecular biotechnology has, however, redefined and reified these technologies as integral to human progress, prosperity, and food security to a new tantalizing level.

Section II comprises Chapters 2 and 3. Of particular interest is the conceptual framework for pharmacogenomics and personalized biomedicine. What constitutes pharmacogenomics and pharmacogenetics? Could these be used interchangeably? What is personalized medicine? This section will also briefly delve into the symbiotic relationships that seem to exist between pharmacogenomics and personalized biomedicine. To recapitulate, many ethical quagmires continue to emanate from these new frontiers and this book elucidates some of the salient ethical issues such as autonomy, informed consent, privacy and confidentiality, genetic essentialism and stigmatization, and genetic tourism. While there are miscellanies of others oscillating on the integrity of the genetic testing and the interpretation of results, accessibility of these tests and associated costs, impact of genomics on the quality of care of patients, *A Guide to Bioethics* probes questions of autonomy (individual and social autonomy), confidentiality, and privacy. The focus of the latter parts of Chapter 2 is on informed consent as a prerequisite in genomic research involving human subjects as individuals and the challenges of group participations in *Genome Wide Studies*. I also discuss genetic essentialism, stigmatization, and genomic tourism. I explore some of the relevant ethical concepts and attempt to apply them to specific topics. For example, what constitutes autonomy? Is there any conflation between individual and social autonomy in terms of genomics and biomedical research? Should physicians override the time tested physician–patient relationship (PPR) and disclose confidential genetic information to third parties without consent? What are the professional and ethico-regulatory guidelines or framework? Are these guidelines absolute? Responses to these nagging questions and conundrums are examined in this section. Genetic services such as gene testing for paternity, ancestral lineages, forensics, disease variants/therapeutic purposes have become a common phenomenon. These services are offered either within or outside of the confines of clinical settings directly to patients herein consumers. The proliferation of this practice has been galvanized with the development and access to the services by many allied and biomedical entities with an insatiable population desirous of these services. In essence, the *Direct-to-Consumer* (GTC) genetic testing model focuses on individuals in providing personalized genetic services and products. Chapter 3 examines the pros and the cons of GTC genetic testing and some of the vexing bioethical challenges.

In Section III (Chapters 4 and 5) some of the regulatory, legal and policy questions vitiating from biotechnology and bioethics are dissected and discussed. For instance, the scientific and regulatory processes in the development of biologics and the formidable roles of the FDA and other local and international agencies in ensuring proper ethical conducts of biomedical research. I will also discuss some biotechnology-based legal issues. Three significant questions are examined, namely, are genes products of nature or are genes human inventions? In addition, this section also examines the Supreme Court's decision on human genes (*Myriad v. AMP*) as a case study since the *Myriad* decision has generated myriads of debates and concerns. I reflect on the aftermath of the Myriad decision. For instance, there have been some assertions following the decision that it may have a chilling effect on innovation especially protein-based or genomic research. Others postulated that as a result of a possible chilling effect, potential investment portfolios in the biotech sector may see a down turn. Were these predictions about the post *Myriad* case demonstratively accurate? On the contrary, there seems to be a popular view that patenting naturally occurring human DNA could derail innovations since it could drive up costs (passed later on to consumers/patients). How are these competing and diametrically opposed views, post the *Myriad* decision? Third, who *owns* genomic materials and information? This book responds to these questions from an interdisciplinary perspective.

In Section IV (Chapters 6 and 7), which constitutes the final part of the book, I excavate some of the challenges in genetic bioengineering tools in the modifications of genes, cluster of genes or entire genomes using nucleases such as *CRISPR Cas 9* and others. It is not uncommon these days to hear of terse discussions about transgenic animals, gene therapy, and micro-gene surgery and the challenges and prospects they seem to present. What are these tools and their applications in biotechnology? Do scientists have justification to create genes or repair mutated genes that could become incorporated into the germline with heightened possibility of being transmitted to the next generation? Could edited genes cause unintended mutations in progenitor cells or genomes? Could these new genes be of any therapeutic relevance to recipients or patients? What about the possibility of aesthetic uses of gene modification tools in designing genes of interests? Could these pose some modicum of harm in stark contradistinction of the ethical dictum of non-maleficence? Are there any ethical guidelines? These main questions are explored in Chapter 6. In Chapter 7, I examine some of the policy trends in recombinant biotechnologies such as gene editing. Some scholars have called for a prudent use of these methods and concurrently vouch for a modicum of paucity or *Moratorium* of gene editing research. It is believed that a *moratorium* will allow scholars, including bioethicists, legal experts, policy makers and concerned individuals and value-based groups as well as funding agencies, the opportunity to examine questions on safety and risks in applying gene editing biotechnologies on humans. On the other hand, some scholars assert that a *moratorium* may become a de facto impediment toward the prudent applications of gene bioengineering for the development of precision medicine. As a result, they argue for a *Noninterventionist* or a *Proactional Approach*. The third emerging approach has been described as *Precautionary*. Proponents of the *Precautionary* approach argue for a middle ground and prudent use of gene editing in precision medicine even though there are genuine questions of safety, but at the same time, the therapeutic prospects remain unprecedented. Rather than a blanket moratorium and unregulated terrain, applications of gene editing tools should be considered apropos and under the expediency of each case until there is proper consensus within the scientific community and professional groups and experts.

The final chapter under the aegis of the title *Perspectives and Conclusions* is a synthesis of the salient bioethical trepidations adduced in the book. Akin to Hegelian dialectics, the chapter summarizes the emerging and nagging issues in biotechnology and bioethics discussed in each chapter. It offers an optimistic and prudent invitation to the erudite reader to ponder over the prospects of biotechnology, its contribution to scientific scholarship, medicine, public health, as well as a "tool" for progress and the development of the next generation of products such as biologics and development. The book makes the case for the therapeutic prospects for personalized medicine, esthetics, human enhancements as well as the meteorites of opportunity inherent in improving almost every facet of socio-cultural life. The choices we make either as individuals or collectively as a society, for example about gene editing, will have some effects on future generations.

A bulk of the questions and issues explored in this book have been partly raised and discussed in the context of my teaching career in bioethics, philosophy, moral philosophy, and the sciences internationally. Students will find the didactic and pedagogical approach to this book helpful in digesting the complex ethical nexus they encounter in their studies and in real life situations. Professionals and students in the biological sciences especially in medicine and allied health sciences, bench scientists, socio-medical experts, policy, regulatory and law experts will find *A Guide to Bioethics*, an unalloyed swath of interdisciplinary scholarship about some of the nascent bioethical quagmires posit by genetics and molecular biology deftly analyzed. *A Guide to Bioethics* is an essential book to every scientist and bioethicists!

Acknowledgments

I am very grateful to Bennet Togbe, MD, Mark Lesney, PhD, and Obiora Anekwe, MEd, EdD, MS, MST, and Emmanuel Sogah, PhD for their invaluable support for their respective critical purviews of the initial draft, and for their insightful criticisms that have greatly shaped the trajectory of the final manuscript of *A Guide to Bioethics.*

I wish to thank my former professor and thesis director during my graduate degree program in bioethics at the Columbia University, Professor Robert Klitzman. Undoubtedly, the thesis forms part of the core of this book. To my doctoral advisor, Professor Chris Emdin, PhD, Teachers College-Columbia University, I am very thankful for the invigorating lectures, his consistent encouragement, support, and challenges in the pursuit of academic and professional excellence. I am thankful to him for his kind words and examples. I wish to express my profound thanks and appreciation to Ms Mindy Asiedu for providing me with the serenity and the support during the initial draft of this book.

I am very grateful to my family for their care and dedication in supporting academic excellence and especially throughout my studies at various colleges internationally. As in the Ewe language, *akpe na mi* (thank you all). To my late parents in particular, Grace and Joseph, I am perpetually grateful for their love and exceptional care, and sacrifice in order for me and my siblings to access excellent education and formation.

Finally, I am grateful to Ms. Jennifer Blaise and Dr. Chuck Crumly at the CRC Press and Taylor & Francis Group, and Mr. Arun Kumar for their persistence, meticulous attention to detail, and expeditious communication even when I had a brief hiatus due to ill-health toward the very last stretch of the manuscript submission. Your unwavering professionalism and encouragement are appreciated and worth emulating.

Author

Emmanuel A. Kornyo holds a teaching licensure in science education from St. Francis College of Education in Ghana. He has also studied philosophy and sociology at St. Paul's Major Seminary, Accra, and theology at St. Peter's Regional Seminary in Cape Coast, respectively. Emmanuel has earned graduate degrees in theology, biotechnology, and bioethics. He is currently at the last stretch of his doctoral studies in science education at Teachers College, Columbia University.

Written by a bioethicist, biotechnologist, and a science educator with extensive teaching experiences internationally, *A Guide to Bioethics* offers a labyrinth of interdisciplinary perspectives on the ethical issues emanating from genetics, personalized medicine, gene editing and its multifarious applications to pharmacogenomics, public health, and regulatory science cognizance of the social contexts of these ethical quagmires. Readers will find *A Guide to Bioethics* refreshingly inviting and tantalizingly elucidatory on the quagmires of biotechnology through the acuity of bioethics!

SECTION I
Bioethics of Biotechnology

1

A Bioethics of Biotechnology

Introductory comment

A captivating scene in the 1997 Academy Award movie, *GATTACA*, portentously suggests that society may be defined by some essentials of genetic traits. The movie illustrates a hodgepodge of clinical tests such as preimplantation screenings aimed at identifying certain desirable genetic traits in order to select some people for specific roles in a "not-too-distant-future" society. Vincent, one of the main characters is born naturally and genetically defective. Tests confirmed that he has bad eyesight, heart problems, and limited life expectancy of 30 years due to his genetic proclivity at birth. He is consequently relegated as an *invalid* and performs menial jobs at the hypothetical space center, *Gattaca Aerospace Corporation*. To avoid these genetic defects, Vincent's brother is meticulously bioengineered and falls under the spectrum of "valids" in the "not-too-distant-future" society. Using an intricate bioinformatics algorithm of machines, members of this futuristic society are easily deciphered/discriminated and categorized into their genetic essentials as *valids* and *invalids*. Workplace discrimination is common and legally permissible and encouraged in accordance with genetic profiles of members of the society. In the meantime, Gattaca Aerospace Corporation is fervently preparing for a space mission to Saturn's Titan. Selection of the mission crew is highly competitive and exclusionary—only *valids* are qualified for the mission. When selected, *valids* will be required to prove their genetic status on a regular basis through biometric tests. Therefore, clearly, Jerome is qualified because he was genetically bioengineered and becomes an astute and accomplished athlete. But there is a hurdle and indeed jeopardy. Jerome had an accident while playing the game of chicken. As a consequence of the injuries, he is unable to participate in the mission despite his excellent genetic makeup and physical prowess. Through an elaborate plot, Vincent assumes the identity of Jerome but is required to pass biometric tests every day throughout the pre-mission training. The movie is punctuated by a stent of a tragic controversy at the mission center. Vincent is cleared and eventually flies the space mission having passed all the biometric tests (though he consistently presents Jerome's genetically enhanced biologics). Critics of the movie have been swift in identifying a medley of ethical issues such as genetic determinism and the question of free will.[1]

Nonetheless, *GATTACA* seems to give some credence to the discussion on the clinical significance of genetics in our time. Indeed, the discovery and affirmation of the structure of DNA as the genetic material and the basis of life and the completion of the *Human Genome Project (HGP)* are irrefutably, among the most important landmarks in biomedical science in contemporary times.[2] Generally, accurate etiological information constitutes some of the fundamental factors in medical diagnosis, prognosis, and the development of therapeutic targets for the treatment of diseases. For example, in the late eighteenth and early nineteenth centuries, the emergence of allied biomedical sciences such as bacteriology, immunology, physiology, pharmacology, and new technologies such as x-rays cascaded in an epistemic shift in improving biomedical diagnosis as well as better treatment of diseases.[3] Currently, advancement in molecular biology, genetics, and the availability of information technologies such as computers and the ability to process and analyze high throughput medical data is paving the way for a new frontier in biomedical sciences and public health. The interests in molecular genetics, in particular, and the sequencing and mapping of the human genome have many widespread applications. Genomic data give credence to the variations and similarities of human beings at the molecular level and this is particularly significant for an array of reasons.[4] Unlike *Gattaca*, genealogical anthropologists could use the genomic information for comparative genotyping in order to construct a global pattern of population trends (including migrations and emigrations) and have a better understanding of specific genealogical loci.[5] In addition, international allied health organizations such as World Health Organization (WHO), American Medical Association (AMA), epidemiologists, and public health experts could also use genomic information in formulating and strategizing interventional measures in order to ameliorate as well as educate the public about certain diseases common to specific locations.[6] Third, data about the human genome are significant in biomedicine for developing diagnostic protocols such as the Breast Cancer susceptibility gene (BRCA) kits, therapeutic and preventive measures such as tailoring precise treatments for patients with specific genetic profiles. For example, the enzyme CYP 450 located in liver microsomes and small intestines has significant roles in the metabolisms of pharmaceuticals.[7] Genotypic expressions of Cytochrome CYP2C9 have effects on a popular coagulant, *Warfarin* dosage, metabolism, and interactions with other xenobiotics.[8] Some studies have demonstrated that patients with genetic polymorphism and mutations in the CYP2D6 gene cannot metabolize some pharmaceuticals such as *Galantamine*, *Donepezil*, and *Rivastigmine*.[9] With such genetic information, pharmacologists will be able to *target* these genetic variations and explore alternative metabolic pathways that will be preponderance to their personalized medical needs.[10]

Another biotechnological feat oscillates on the genetic modification of nonhuman organisms and products. Of particular interest is GMO rice! Vitamin A is significant for vision, immunity, and growth and in the improvement of general health. Vitamin A deficiency (VAD) causes blindness and other pernicious health problems and incidents of childhood mortality. The WHO estimates that over one million children, mostly in relatively underdeveloped or emerging economies, are predisposed to VAD. Global initiatives have failed to eradicate this menace of the VAD. However, biotechnology seems to have a pragmatic solution! Rice is a global staple food, and biotechnologists, public health, and other experts have pondered and proposed biofortifying with Vitamin A. Therefore, through the process of bioengineering, biotechnologists with international support have successfully designed new rice, herein *Golden Rice 2.* Generally, rice produces beta-carotene in its leaves during development but the two genes are naturally switched off during grain production. As a result, grains of rice produced by conventional methods lack the desirable beta-carotene. The two genes have been biotechnologically inserted to bolster the production of the beta-carotene. Golden Rice 2 is biofortified with beta-carotene, a precursor to Vitamin A. Extensive research and field tests show that the genetically modified rice has expressed beta-carotene, determined to be bioavailable, and safe for human consumption. It is anticipated that if produced and consumed on a large scale within the target population, it will invariably help assuage the menace of VAD without a resort to medical and public health interventions. However, the notion that Golden Rice was a GMO has caused unprecedented concerns both within academia and the public. Some of the concerns are safety, potential allergenicity, bioequivalence, and the possibility of crossbreeding with the local rice although there is no scientific evidence to substantiate these claims. Nonetheless, these perceptions have become the epicenter of repugnancy and oppositional platform for segment of the public about the "dangers" of biotechnology despite the golden opportunity GMOs seems to offer.[11]

The examples enunciated above seem to demonstrate the potentials that the bioengineering of genes could have on individuals and on society as a whole. In brief, the completion of the *HGP* has given a new thrust in the development and practice of medicine, public health, and others.[12] This piece attempts to explore the significance of the sequencing of the human genome and in particular single-nucleotide polymorphisms (SNPs) and genetic engineering in defining the new frontiers in medicine such as pharmacogenomics and personalized medicine (PM).[13] But these corpus of scientific advancements (as anticipated) has generated ethical and socio-policy concerns. In fact the *HGP* anticipated it, and so it provided detailed ethical framework called the *Ethical, Legal and Social Implications (ELSI) Research Program* under the auspices of The National Human Genome Research Institute (NHGRI) for guidance. In addition, in order to guide researchers the United States' Presidential Committee for the Study of Bioethical Issues, the Nuffield Council on Bioethics (UK) among others have also made significant inputs in this enterprise.[14] The *ELSI* in particular addressed issues about privacy, fairness, research subjects, the integration of genetic technologies, the education of healthcare professionals, policymakers, and the public on the implications of the genetic information among others.[15] But biotechnology as a scientific discipline has deep historic roots transcending almost every known epochs. Therefore, I attempt contextualizing biotechnological innovations historically. In addition, I examine the nature of biotechnology as a scientific discipline. I believe these two perspectives will lay the foundation for a better bioethical analysis of biotechnology.

The biotechnology of history

One of the prolific writers of all times Cicero, purportedly said that "the causes of events are ever more interesting than the events themselves." This aphorism (in my humble submission) seems aptly applicable to the discourses and the confluence about biotechnology and history. There are several historical accounts about the emergence of science and in this context biotechnology since antiquity. Scientific discoveries as we have them in most historical books appear to focus on the "events" oscillating on some of the great feats about and of scientists and their impact on humanity or the world. Some historical accounts tend to present science as purposeful, well-orchestrated, and orderly "discoveries" of great men and women. Of course, there is also the notion of the critical purviews of the history of science in which these are emphasized as well as contexts and other issues surrounding some scientific feats are additionally scrutinized. It is in this later approach that the erudite reader discovers the apparent messiness, challenges, and sometimes the very true nature of how and why specific scientific ideas, concepts, and hypotheses work in favor of scientists leading up to some of the discoveries accounted in our historical books. In Ciceriam terms, I believe the *causes of events* of science especially biotechnology is perhaps more interesting than the conclusions drawn about the history of biotechnology. In other words, as scientists, we should be pondering on the causes of the initial and perhaps crude biotechnologies used since antiquity until contemporary times. Why did people across almost every epoch use some form of biotechnology in enhancing the qualities of their lives? Why do we still use yeast in making bread, universally? Why and what was the event that prompted ancient people and modern humans to use the process of fermentation? What events in history galvanized the need for the domestication of crops such as potatoes and wheat? Are these events in history recurring and why do modern men still strive to improve crops even after at least 10,000 years of domestication of some of these same crops? For instance, rice was reportedly domesticated in parts of Asia (Pearl River Valley) over 10,000 years ago. Today, Golden Rice, which is beta-carotene enriched has taken stage in the field of biotechnology with mix of positive and negative perspectives among a segment of the population.[16] The use of antibiotics in medicine transcended many annals in history but in our time, it is generating concerns. In brief, history has accounts of many scientific events that seem cyclical.[17]

In view of the above, my intent is not to regurgitate historical accounts of biotechnology but rather to explore the *biotechnology of history* that reflects on the factors undergirding the emergence, development and applications of biotechnological methods as a phenomenon within human history. My focus in this chapter among others is not to offer a chronological reflection as such, but attempt to highlight some of the biotechnological methods such as fermentation, hybridization, domestication of crops and animals, protein, and recombinant DNA (rDNA) biotechnologies. I believe this will lay the foundation for an ethical reflection and the need for regulating some of the innovations and the processes of biotechnology.

Undoubtedly, biotechnology products have always been around us. Indeed, biotechnology *products* have become coterminous with contemporary life: from bread, beer, cheese, dyes, biologics (biopharmaceuticals), biofuels, genetically modified organisms (GMOs—plants and animals), and reproductive and regenerative technologies just to enunciate a few. The processes for developing biotechnology have been valuable as its products. Some of the processes of biotechnology include fermentation, plant and animal hybridizations, and preservation methods such as drying among others. Biotechnology is increasingly becoming a household name and a popular genre in the field of biological sciences. It remains a bourgeoning but active academic endeavor with significant economic, social, financial, political, and ethical implication of global proportions. Nevertheless, it has also generated copious trepidations in law, policy, in the biomedical field, and among a segment of society. Put succinctly, there seem to be an aura of suspicion around the discovery process, the subject matter, and the products of biotechnology despite the good aspects of its products on humanity evidenced in every known documented history. Unlike other subject matters that have true critical historic accounts, biotechnology has some symbiosis on human activity in history to the extent that it appears subsumed in the annals of historical accounts. In other words, there are virtually no extant discussions of biotechnology as a genre in the history of science as such. Given this lacuna, this piece attempts at summarizing the biotechnology of history rather than the history of biotechnology as there are no primary data per se in elucidating when, how, and why biotechnology emerged as a true scientific discipline. I will offer a definitional exposition about the concept of biotechnology, when it was first used and why, as well as a tacit but operational description in contemporary times especially in the contexts of molecular biology and genetics, in particular.

Therefore, the term biotechnology encompasses several facets within academia. Biotechnology uses technology with living systems such as cells, tissues, or any biologic to create novel products or assays. According to the US Office of Technology Assessment, biotechnology entails "any technique that uses living organisms or their products to make or modify a product, to improve plants or animals, or to develop microorganisms for specific uses." Biotechnology covers a broad frontier of the application of the physical sciences such as engineering, mathematics, and technology to biologic systems such as cells and tissues to generate products. Biotechnology methods are applicable in the production of biopharmaceuticals; genetic modifications/bioengineering of organisms ranging from viruses, bacteria chromosomes with wide range applications for human and nonhuman use. Here the terms biotechnology and bioengineering are used interchangeably due to their stark similarities. Nevertheless, what is biotechnology and how does it differ from applied biological science? What is the origin of the concept and when did it get into the scientific lexicon? The word biotechnology appeared in the genre and history of science relatively recently in the twentieth century. An engineer from Hungary by name Karl Ereky is generally reputed to have coined the term biotechnology. He was an agricultural engineer and an entrepreneur. In a book or manifesto published in 1919 entitled, *Biotechnologie Der Fleisch-, Fett- Und Milcherzeugung Im Landwirtschaftlichen Grosbetriebe: Fur Naturwissenschaftlich Gebildete Landwirte Verfasst (1919)* (i.e., *Biotechnology of Meat, Fat, and Milk Production in Large Scale Agricultural Enterprise*) he described biotechnology as "all lines of work by which products are produced from raw materials with the aid of living things." The living systems include yeast, bacteria, and other microbial organisms used in the fermentation and *transformation* or production of large-scale agricultural products in the twentieth century. Akin to the industrial revolution, Ereky postulated the thesis for the optimization of technologies with biological systems in the production of food for an increasingly insatiable population of his time. He believed biotechnology could resolve some of the dire challenges of his time such as food security and energy production. He demonstrated this by investing and raising many farm animals such as pigs on a large scale among others.[18]

Furthermore, the word "biotechnology" first appeared as a *hapax legomenon* in the English Language as a title to an article in the April 1933 issue of *Nature*—although the word was not given any detailed exposition in the actual piece. In 1938, Julian Huxley suggested, "Biology is as important as the sciences of lifeless matter, and *biotechnology* will in the long run be more important than mechanical and chemical engineering."[19] Evidently, the concept of biotechnology gained full academic usage in the twentieth century. Indeed, the three descriptions and contexts in which the term was used foresightedly suggest that biotechnology will be more significant than other scientific subjects at their time and in the future. Such teleological or futuristic perspectives may be considered pioneering as they seem to identify the usefulness of biotechnology in almost every facet of human life, especially the diverse products, to meet huge societal demands. It is worth noting that despite the relatively recent appearance and usage of the term biotechnology, humans in many epochs (spanning over 10,000 years), have actually used various biotechnology methods in the transformation of raw materials such as wheat and yeast to produce bread. It may be therefore anachronistic to suggest that the term biotechnology as used today has always been part of early human scientific semantics and history. While it is true that biotechnological activity preceded the coinage and usage of the term "biotechnology," suffice it to say the products have always been part of human civilizations unabated.

Generally, scientific historians and archaeologists postulate the thesis that some of the earliest forms of biotechnology emerged shortly after humans began to transform their lifestyles from nomadic to sedentary, 10,000 years ago. As early humans moved toward sedentary lifestyles, crops and animals with desirable traits were meticulously selected. These plants and animals were cultivated and later domesticated for improved yields. Even though the ancient biotechnological forms appeared simple, it propelled and sustained their sedentary and aggregated lives. It also led to unintended transfer of desirable "genes" from one living entity to another without knowing the details of the molecular basis for their ventures as known today.[20]

As humans settled in small herds of populations around caves, they needed to secure a constant supply of food and other essentials for their lifestyle. Early humans selected crops and animals that they could feed on and use for fuels, clothing, and other purposes to sustain their lives. Therefore, some of the earliest forms of biotechnology were the domestication of plants and animals to sustain sedentary populations. Dogs, goats, and cattle in particular were bred under the expediency of many useful purposes such as their size and good features for hunting, grazing, and protection over 10,000 years ago. Later, dogs and other domesticated animals were selectively bred or crossbred for improved characteristics that served many purposes for ancient humans. Ancient humans such as Egyptians, Assyrians, Aztecs, and Babylonians also domesticated plants, such as wheat, barley, and corn. In sub-Sahara Africa, sorghum was the most domesticated crop to support huge populations. Similar to animals, plants were crossbred for food and medicinal purposes. For instance, several forms of wild corn were reputed to have been crossbred with domesticated ones leading up to improved breeds for disease-resistance, better yields, and other traits. In the corpus of modern biotechnology jargon, we could say, these ancient practices of domestications were forms of hybridization! The ancient people also used salt as a preservative for meat and developed highly sophisticated embalmment procedures.

In medicine, biotechnology in its fundamental forms have been used and documented among some ancient peoples as in China, Egypt, Aztec, and others. Egyptians, for example, harvested honey and used it to treat respiratory infections as well as in treating wounds. It is fascinating to note that honey contains a natural antibiotic; therefore, ancient Egyptians used it to dress wounds, ostensibly to curb microbial infections, and to prevent further infections. The ancient Chinese also made insecticides from the chrysanthemum plant. In addition, ancient Chinese used moldy soybean curds (which contained antibiotics) to dress and treat wounds. Of course, later, Fleming will identify and isolate penicillin for medicinal uses giving vent to a robust modern antibiotic industry with significance for medicine and obviously for the improvement of the quality of human lives.[21]

In addition, many ancient cultures, such as Chinese Egyptians, and Greeks, practiced some form of zymotechnology—which in essence is the use of yeast or bacteria through the process of fermentation to make cheese, yoghurt, wine, beer, bread, among others. Fermentation is one of the main processes used to transform raw materials into products even in contemporary biotechnology practices. These practices have transcended several millennia to our time. Fermentation is still used in upstream bioprocessing to produce biologics, beer, yogurt, and an avalanche of other products.[22] Egyptians added microbes such as yeast to fruit juices to produce wine and also vinegar for many uses. This technology became popular among a segment of the ancient near-Eastern people such as the Sumerians, Assyrians, Israelites, and Egyptians spanning many generations as the technology was easily transferrable. Several ancient texts such as Ancient Near East Text (ANET) have referred to the use of wine, vinegar, and bread in these cultures. Another application of biotechnology was in the use of dyes in textiles, paintings, and pens.[23] Until the discovery of artificial dyes in the mid-1900s, most dyes were extracted from the bark of trees. Once the desirable barks were harvested, they were added to the fabric and boiled. This simple biotechnology process, which allows the transfer of the colors from the bark to the fabric, is an example of one of the enduring forms of biotechnology. Whereas some of the ancient Middle Eastern peoples such as Egyptians "extracted" papyrus and produced crude paper, the ancient Chinese developed a highly sophisticated method of producing paper. In other words, the application of some forms of technology to biologic systems has become socioculturally entrenched and normalized in ancient societies and tacitly transmitted to us. Indeed, by the twentieth century, fermentation processes were optimized in the production of organic compounds such as acetone, ethanol, and other biofuels. Chaim Weizmann's pioneering work on the process of glucose fermentation was developed and optimized to industrial levels to produce cordite, a formidable product in World War II. The large-scale production laid the bedrock for subsequent industrial revolution of the postwar twentieth century industries in Europe and in the United States. The discovery of proteins and the isolation of enzymes have been a game changer in the annals of biotechnology by the early nineteenth century. Louis Pasteur used microbes in fermentation in 1857, leading up to the formulation of the germ theory. Two years later, Charles Darwin's seminal work on the theories of natural selection and evolution were published. These and subsequent Darwinian theories have formed the backbone of molecular biology. By the nineteenth century, Gregor Mendel, an Austrian monk discovered the gene to be the inheritable basis for life. Later, James Watson and Francis Crick worked on the structure of DNA—an important breakthrough in science as Mendel research and publications have laid the foundation for modern recombinant biotechnology (rDNA). The field of vaccination also marked an important milestone in biotechnology. Edward Jenner successfully inoculated a child with the small pox vaccines. Alexander Fleming also discovers penicillin which becomes popular in the early and later part of the twentieth century, as many corporations such as Pfizer spend substantial resources into the mass production of the antibiotic penicillin and other vaccines at this time. These seminal works and later research will pave the way for mass vaccination leading up to the eradication of many debilitating infectious diseases. Between the years 1930 and 1945, both MIT and UCLA had departments specifically for biotechnological research and development. Also, Henry Wallace created the first biotechnology hybrid corn for commercial purposes at Pioneer Seeds (Des Moines) and DNA was isolated in a test tube by 1958, though Friedrich Miescher had in 1869 isolated and identified DNA (nuclein) in living systems. He noted in one of his correspondence: *In my experiments with low alkaline liquids, precipitates formed in the solutions after neutralization that could not be dissolved in water, acetic acid, highly diluted hydrochloric acid or in a salt solution, and therefore do not belong to any known type of protein*. This nuclein, was later and properly identified as DNA, which forms the basis of life in living systems. Other scientists such as Phoebus Levene and Erwin Chargaff (1935) also conducted extensive studies on the nature and structure of DNA, which led to the Chargaff rule now in biochemistry and greatly helped in determining the structure of DNA. Oswald Avery, Collin McLeod, and Mclyn McCarty (1944) identified DNA as the repertoire of genetic information that is also transmittable. However, Watson and Crick's model of a double helix three-dimensional (3D) figure of the DNA was eventually accepted. Indeed, "for their discoveries concerning the molecular structure of nucleic acids and its significance for information transfer in living material" Watson, Crick, and Maurice Wilkins were jointly awarded the Nobel Prize in Physiology or Medicine (1962). Fred Sanger successfully sequenced DNAs, which also has been an important breakthrough leading to the HGP and explosion of the next generation of genetic diagnostics tools. Robert Holley, Marshall Nirenberg, and Gobind Khorana were jointly awarded the Nobel Prize in Physiology or Medicine for their "interpretation of the genetic code and its function in protein synthesis" (1968). They *cracked* the genetic code to be three codons that formed the 20 amino acids. The chemical compositions, structure, and various components of the DNA and the RNA were also known laying the foundation for modern recombinant biotechnology and molecular biology.

Furthermore, the discovery of restriction enzymes in 1970 also opened a new chapter in biotechnology. Restriction enzymes are nucleases that cut DNA at specific places and allow scientists to manipulate genes of interest for an array of reasons. Restriction enzymes allow the bioengineering of biologic systems. Restriction enzymes became a molecular arsenal to open and manipulate nucleic acids as desired for innumerable reasons for biotechnologists. It helped scientists to identify genes of interests, better understanding of genes, or cluster of proteins. Restriction enzymes may be compared to cleaning a building structure of all debris and having the clarity and the ability to unpack the materials, specifically the blocks carefully at specific points without destroying them; and having the ability to systematically reposition the blocks and, resealing them after studying. Thus, the concept of "engineering" a biologic system became synonymous with biotechnology. This is because restriction enzymes allowed the bioengineering of living systems and the development of new biologics and products. Literally, scientists were able to cut and paste DNA chunks of interest, to suit specific purposes and develop new or similar products. For example, Stanley Cohen and Herbert Boyer bioengineered DNAs with restriction enzymes. This marked an important milestone in recombinant biotechnology. This is because, for the first time, biotechnologists were able to engineer genes of interest in other living systems as hosts, and harvested and purified these products for many purposes. This is revolutionary and decisive! Biotechnology tools have brought about significant changes in the biological disciplines—biochemistry, genetics, molecular biology, genomics, and many more. The 1970s may be described as an era in which scientists had almost all the biologic tools in a *Pandora box* and the seeming unrestrictive tools at their disposal. Transgenic animals and plants became possible toward the end of the 1970s. In the meantime, astronauts landed in space. Mega computers were also emerging to process and store information from most of these scientific discoveries. There seemed to be a scientific fervor globally leading up to enormous competitions unprecedented since the end of World War II. Concerns were brewing about the possibility of bioengineering complete living entities such as transhumans or animals vividly captured in Halden's classic essay, *Daedalus* (Science and the Future). The concerns transcended the scientific field and the larger society. Also, the horrors of the Tuskegee Syphilis Project had taken center stage as many of the actual research procedures and findings including the roles of the agencies in perpetrating the horrific but surreptitious treatment of innocent black men under the expediency of scientific research emerged and became public knowledge. These and other concerns led to the *Asilomar Conference*, which provided an opportunity in which scientists and other professionals, such as lawyers, could pause, reflect, discuss, and decide on many of the emerging tools and researches in biotechnology. Of significance were the future and direction of gene editing tools. This was because the discovery of endonucleases (which can cut DNA at specific recognition sites) use to edit genes or entire genomes even in humans.

By 1981, the Indian born microbiologist, Prof. Ananda Chakrabarty, successfully bioengineered the *Burkholderia cepacia* bacterium, optimizing its ability to digest oil. The bacterium was used successfully for bioremediation and was patented in the United States. The patent was challenged, but the Appeals Court determined the validity of the patent averring that though the bacterium was a living entity, "the relevant distinction is not between living and inanimate things" but those of "human ingenuity and research" because the bacterium was bioengineered in the laboratory, hence the process and the product qualified as intellectual property. The ruling ultimately served as a legal template in patenting transgenic animals and genetically related product marking an important milestone in biotechnology. It also served as catalyst for researches involving the manipulation of genes and the emerging of biotechnology post the Asilomar Conference and the Belmont Report.[24]

In addition, the discovery of the polymerase chain reaction (PCR) also consolidated the advancements in biotechnology to some extent. The discovery itself involved several collaborations among scientists and the reliance on the extensive dossiers on the nature, structure, function, replication, and sequencing of DNAs. A symphony of these helped Mullis and his team determine the process to amplify DNA to any desirable number. Currently, the PCR is used in diagnostics, forensics, and as a research tool in biotechnology.

Gradually, the frontiers and the process of biotechnology focused on the development of biopharmaceuticals for human consumption. rDNA biotechnologies had established rigor, clarity, and the issues of safety of products regulated internationally. These paved the way for the development and approval of the first recombinant vaccine. The inactivated hepatitis B vaccine,

Heptavax, was developed by Merck Pharmaceuticals and was approved by the FDA for use in humans (though it was discontinued in 1990 in the United States). Human insulin became the first rDNA synthesized biologics/biopharmaceutical for approval in 1991. Several rDNA drugs in R&D have been approved for human use as alternative to chemically synthesized pharmaceuticals. Obviously, the development of biologics was a game changer in medicine and the public health sector. As these first generation of biologics are experiencing *patent cliffs*, a new wave of biosimilars are also emerging with the anticipation that the cost associated with innovative or referenced biologics may drastically reduce in order for them to be accessible to patients. But these laid the foundation for the completion of the HGP in the United States.

In essence, the *HGP* is one of the most internationally collaborative and successful work in the annals of biotechnology. In 1990, the National Institutes of Health (NIH) and the Department of Energy came out with an initial five-year plan on sequencing and mapping the entire human genome. The initial project, titled *Understanding Our Genetic Inheritance: The Human Genome Project, The First Five Years, FY 1991–1995* was under the directorship of Watson and Crick (who later resigned). As a kind of prelude to the HGP, the NIH sequenced and mapped four microorganisms, namely, *Saccharomyces cerevisiae*, *Caenorhabditis elegans*, *Mycoplasma capricolum*, and *Escherichia coli*. On October 1, 1990, the NIH in conjunction with the U.S. Department of Energy officially launched the HGP to determine the 3bn human DNA basepairs and sequencing of the entire human genome and creating a database to restore these information! This was an ambitious and mammoth project to understand the entire human genome. As Francis Collins later noted with respect to the focus of the HGP, "Building detailed genetic and physical maps, developing better, cheaper and faster technologies for handling DNA, and mapping and sequencing the more modest-sized genomes of model organisms were all critical stepping stones on the path to initiating the large-scale sequencing of the human genome." Several scientists and experts around the world—France, Britain, Australia, China, Japan, among others—collaborated and jointly led the research. By 2002, the *First Draft* was released earlier than expected. The *Final Draft of the HGP*, considered the full report was presented by Dr. Francis Collins in 2004. The completion of the human project has paved the way for a new fervor in molecular biology, medicine, genomics, and pharmacogenomics. While the HGP was ongoing, several notable biotechnological innovations and activities occurred in tandem. Monsanto's GMO tomatoes were approved for human consumption in 1997 amidst a global controversy over safety. Dolly was cloned, 1998, Human Embryonic Cell lines were established, and in 2004 CopyCat was cloned for its owner. Biotechnology has transcended every aspect of human life—medicine, industrialization especially upstream and downstream methods have advanced and used as a process in R&D in the biopharmaceutical sectors; a move toward precision medicine and patient care, biofortification of food just to enunciate a few.

In brief, modern biotechnology therefore encompasses a wide range of applications such as DNA sequencing, forensics, cell-based assays, diagnostics assays, tissue engineering, fermentation and bioprocessing, nanotechnology, vaccinations, regenerative medicine (3D scaffolding to regenerate bones, tissues for replacements or for modeling *in vivo* systems), and biopharmaceuticals due to insatiable global demands for biologics and biosimilars. Biotechnology methods have been used to produce and generate alternative energy—specifically the production of biofuels, ethanol from corn, soybeans, and sugarcane. It is also useful in bioremediation and in the production of biodegradable products such as plastics. Biotechnology, especially rDNA, has changed the trajectory in crop improvements and the global food supply chain. Genetically bioengineered food has been on the ascendency especially in the past three decades. GMOs include corn, wheat, beta-enriched rice, and yeast (for the winery, cheese industries). According to the *Global Biotechnology: Market Research Report*, IBIS suggests that the biotechnology financial output was valued at $350 billion in 2016, predicted to grow at the rate of 3.6% for 2016 financial year.[25] Thus, the biotechnology enterprise remains a viable sector of local and global economic significance. There is no doubt that biotechnology has contributed significantly to improving almost every facet of human life: food, fuel production, medicine, public health, clothing, bioremediation, research and development, and forensics just to mention a few.

Despite these applications and uses of biotechnology, axiomatically, there seem to be a general lack of appreciation for the sector and its countless innovative products. In particular, there have been many public outcries and condemnations of biotechnology products such as GMO crops and animals. Increasingly, several publicities including adverts, seminars, and discussions are aimed at discrediting the significance of biotechnology in the world. These scenarios and apparent lack of appreciation of biotechnology science and products posits serious quagmires and questions. As Troy Sadler noted, "Despite the significance of biotechnology within the sciences, it has not become a prominent trend in science education….the ideas, tools and products of biotech are transforming science and society (including production of food, treatment and diagnosis of disease, manipulation of genomes, changes in the workforce needs), but these developments remains vastly under-represented in the curricula and classrooms."[26] Is biotechnology bad per se? Are biotechnologists communicating their findings and products properly to the public? What is the level of biotechnology scientific literacy among the general population? Is biotechnology being taught in our schools at all? What is the component of the biotechnology curriculum? What indeed undergirds this global oppositional phenomenon to the genre of biotechnology? Is there anything in the nature of science that helps elucidate these challenges? Or does the *Nature of Science (NOS)* contribute to these apparent misconceptions and controversies oscillating on biotechnology?

In perspective, this section on the biotechnology of history has identified and highlighted some of the causes and events of biotechnology from ancient to contemporary times. Every epoch and moments of man/woman have consistently demonstrated their scientific and dynamic ingenuity to be futuristic and ensure the basics of life such as food security, medicine, and economics by improving upon the status quo with science and in particular, biotechnology even though, the term biotechnology was axiomatically coined relatively recently. A discovery in biotechnology is often an aggregation on past scientific discoveries and tools. By applying concepts, ideas, tools, and technologies to living systems, biotechnologists strive to generate new concepts and products. As a scientific discipline, biotechnology reflects the very essence and nature of the scientific and the scientific trajectory. But how do biotechnologists operate? How do they undertake research? Do they adhere to certain guidelines when conducting research involving living systems? What kind of science and technology is in biotechnology? To respond to these and an avalanche of questions, the second portion of this chapter offers a reflection on the nature of science and how bioethics emerged. While scientists generally adhere to certain ethics and codes of conducts, the genre of biotechnology has provided a unique set of challenges. This is evidenced in the emergence of federal and international ethical documents that has played a significant role in shaping bioethics and the regulation of biotechnology.

The nature of science and biotechnology

A research paper, "*Spontaneous Human Adult Stem Cell Transformation*," initially considered to be a breakthrough, was retracted to the chagrin and delight of many scientists. The publishers indicated in their retraction editorial, "The authors retract the article titled 'Spontaneous Human Adult Stem Cell Transformation,'" which was published in the April 15, 2005, issue of *Cancer Research*. Upon review of the data published in this article, the authors have been unable to reproduce some of the reported spontaneous transformation events and suspect the phenomenon is due to a cross-contamination artifact. Five of the seven authors have agreed to "the retraction of this paper." As the editors have noted, the authors, other scientists, and peers could not replicate the findings even though the paper was well cited (at 300 times) before the retraction. But was it the intention of the authors to mislead the scientific community about their findings? Why and how could they not have been careful in "following" standard scientific protocol in detecting and precluding the cross-contamination of the cell lines? While these are legitimate questions, nevertheless, the frontiers of science have and will continue to experience these sorts of challenges. This is because science, by its very nature, is a process and a dynamic enterprise! As a dynamic enterprise, it has its own gamut of methodology in generating its own corpus of knowledge even though not every form of knowledge is scientific in scope. The integrity of scientific knowledge is often further reified, refined, and subject to the crucible of some rigor in order for it to qualify as science. Sometimes, the presumptive facts of science may metamorphous due to the generation of new or better interpretation of data either to validate or refute some theory or scientific facts. Hence, what may be considered normal science may actually be ethereal throughout the course of the history of science.

This is evident as several ideas, concepts, and discoveries in biotechnology have undergone some seismic and paradigm shifts over the years due to new insights, perspectives, and a conglomeration of other factors. Several scholars including Kuhn have attempted to clarify this amorphousness when he poignantly stated that "normal science" is a *research firmly based upon one or more past scientific achievements, achievements that some particular scientific community acknowledges for a time as supplying the foundation for its further practice.* Normal science thus has a research and a methodology that is firmly rooted in some *scientific precedents* evidenced in actual achievements that are *validated* by a scientific community or experts. In other words, science has a social context, that is, scientific achievements occur within specific communities of people and indeed in the words of Kuhn, "attract an enduring group of followers." Therefore, contrary to the presumption that science is an "objectively" close-knit or absolute field of knowledge, the processes of normal science and discoveries are, however, markedly open-ended frontiers, with unresolved questions for its followers (scientific community) to resolve in the context of their own historic and social milieu accompanied by new data and interpretation. Current scientific knowledge is symptomatic attempts to resolve some of these pending issues within their respective scientific communities based on current data and interpretation. In other words, normal science provides the framework for continuous scientific inquiry and dialogue in every historic epoch. In addition, science is also a sociocultural phenomenon confined to a specific geophysical milieu and place. As such, it is a social enterprise comprising communities of scholars within the discipline with implications for the larger social community of policymakers, educators, physicians, politicians, and many others. To put it subtly, science is a social construct in which the knowledge, skills, and technologies generated are generally construed to be coproduced. The coproduction of scientific knowledge therefore could be in the halls of academia, a laboratory, or in specific places such as in a biotechnology company. Undoubtedly, society is curious and interested in the technology and products that proceed from the processes of the scientific enterprise. Historically, society has played a significant role in shaping the scope, contents, and directions of scientific innovations by actively supporting the process of research and discovery through the infusions of financial and human capital. Thus, the scientific process is shaped by societal criticisms—sometimes constructive and other concerns. Others include peer reviews in scientific papers, conferences, replications of scientific procedures, and others. Sometimes, scientists or companies may recant their hitherto ideas or recall products they find problematic for the public good. For example, on April 21, 2015, Sanofi Pasteur Swiftware, a pharmaceutical company, voluntarily recalled its flu vaccines. In an official statement to the FDA, the company stated, "As part of Sanofi Pasteur's ongoing monitoring of the stability of all their influenza vaccines, they found that the antigen content of 3 lots of the 2014–2015 FluzoneQuadrivalent vaccine supplied in multidose vials has declined below the stability specification limit for 2 strains—A/Texas H3N2 and B/Brisbane (Victoria lineage)." These seemingly routine occurrences make the case for the regulation of science by the coproducers of the scientific knowledge and products they find problematic or at variance with socio-scientific and safety conversions.

Furthermore, certain discoveries may be serendipitous. Indeed, many significant scientific discoveries and or innovations emanated from what might have been hitherto considered "anomalies" and serendipities. For instance, x-rays were accidentally discovered during what was considered a *normal scientific experiment* on cathode rays in which Roentgen observed that a barium platinocyanide screen emitted light or glowed when it should not under *normal experimental conditions.* Such serendipitous discovery led him to investigate the phenomenon that has led to the actual discovery of x-rays with profound clinical applications.[27] In fact, Rosalind Franklin's x-ray crystallographic data played a significant role in the discovery and affirmation of the 3D structure of DNA postulated by Watson and Crick. Of course, the nature and structure of DNA and RNA lies at the very epicenter of biotechnology! Another classic example (in the field of biotechnology) was the discovery of Viagra—a buck buster biopharmaceutical generating over $1 billion in 2015 alone, which was initially meant to treat a heart condition called angina. However, during clinical trials, investigators discovered that Viagra was not going to be clinically potent for the treatment for the purported clinical indication. Fortuitously though, some male research subjects reported some *unusual symptoms* of increased erections or *anomalies,* which has become synonymous with the use of Viagra now. As Kuhn noted "...awareness of anomaly plays a role in the emergence of new sorts of phenomena, it should surprise no one that a similar but more profound awareness is prerequisite to all acceptable changes of theory."[28] Another example worth the discussion is penicillin, which was serendipitously discovered by Fleming in 1928.[29] Other accidental discoveries include pacemakers and the pap smear! As we will soon see in this chapter, the scientific process and, for that matter, biotechnology may be analogous to a swamp in which every miniscule of its content has a potential for leading to a discovery and a product.

It is also worth noting that biotechnology and the scientific enterprise like any other area of academia operates through the apertures of social conventions, policies, and regulatory norms typically mandated to ensure the safety of scientific products for consumers. Science must be socially relevant. It must enervate the human condition and improve the quality of life of its citizenry. It must bring about some form of social transformation and address peculiar challenges. However, certain sociohistoric incidents gave heightened sense of regulatory and legal frameworks to regulate the process of science especially in the arena of biologics and genetics in particular. Despite the significant roles society exerts in the contemporary scientific process, suffice it to say, it has not always been the case in the past. Indeed, the scientific community has been besieged by pseudosciences, quackery, and wanton greed that once put the larger social community at risk, historically evidenced in the annals of science history. For example, many quack doctors "believed" that smoking *cigars cures asthma.* Today, scientific evidence and research actually shows the opposite effects of smoking. Of course, the argument has also been that science is self-regulatory and there is a truism in this assertion too.

In perspective, the notion that science is a process is demonstrably true.[30] Nevertheless, the scientific enterprise occurs within specific sociocultural purviews. As a result, science and the process of scientific innovations has been regulated universally especially at the beginning of the twentieth to the twenty-first centuries in particular. Regulatory bodies such as the FDA, European Medical Association (EMA) (formerly the European Agency for the Evaluation of Medicinal Products—EMEA), and Health Canada, to mention a few, have been foundational in the regulation of science and the protection of public health. Together with intellectual property and case laws, the scientific processes have been regulated and several issues adjudicated in the courts of law, by experts within the community, and policy experts among others. So, what does science regulate and why? Is it necessary? Could regulation truncate scientific and therefore biotechnology innovations? Can society trust biotechnologists?

The scope, the nature of science, and biotechnology continue to be at the fulcrum of the advancements of humans transcending almost every facet and epoch of civilization. Scientific inquiries continue to generate copious amount of knowledge and discoveries even at its rudimentary levels. Of particular focus is the output and regulation of biotechnology. There is no doubt that biotechnology has seen a surge in the corpus of academia and industry especially in the past four decades. Biotechnological innovations have significance for the socioeconomic, moral, political, cultural, and even philosophical life of almost every human being. I believe bioethics has become the buffer zone in discussing and analyzing these conundrums; hence the book, *A Guide to Bioethics*.

What then is the bioethics of biotechnology?

The term bioethics did not enter the lexicon of scholarship until recently, considering the fact that "ethics" had already established tradition in philosophical discourses over two millennia. However, "bioethical" issues have always been in existence. Etymologically, the term comes from two Greek words "bio" and "ethike" or "ethos." The former means life, while the latter implies habit or custom. The juxtaposition of these two words is somewhat encapsulated in the Webster's Dictionary definition as *a discipline dealing with the ethical implications of biological research and applications especially in medicine*. The word "bioethics" or specifically, "bio-ethics" appeared in an article written by Fritz Jarh titled, *Bio-Ethics: A Review of the Ethical Relationships of Humans to Animals and Plants*. He used the Kantian categorical principle in making the case for better attitudes and obligation towards non-human living beings as well as human beings. However, there is another argument that Prof. Van Renselaer Potter coined the term "bioethics" as used and understood in current scholarship. Potter saw bioethics as *...a discipline which combines biological knowledge with a knowledge of human value systems, which would build a bridge between the sciences and the humanities....*[31] For Onora O'Neil, bioethics is *...a meeting ground for a number of disciplines, discourses, and organizations concerned with ethical, legal, and social questions raised by advances in medicine,* *science and biotechnology.*[32] These two definitions clearly bear the cloak of the multidisciplinary characteristics of bioethics. This is because bioethics focuses on "living systems" by applying broad concepts drawn from philosophical ethics, sociology, law, anthropology, and history to analyze and offer discussions and perspectives on some trepidations often encountered in biotechnology or biological sciences. Bioethics has become part of the corpus of the genre, applied ethics that addresses nagging and broad range of concerns emanating from the biological sciences, while biomedical ethics narrowly focuses on ethical issues in the biomedical fields in particular. The two terms, bioethics and biomedical ethics may be used interchangeably. When used in this book, bioethics implies ethical matters broadly, unless narrowly and contextually defined. Hence this book, *Bioethics of Biotechnology* offers an interdisciplinary exposure on the challenges posed by contemporary biotechnological researches and applications through the acuity of ethics. In fact, biotechnological science will be analyzed through the aperture of concepts and norms from philosophical ethics, epistemology, regulatory and policy, law, history, social anthropology, and international conventions and norms in order to draw bioethical conclusions.

As intimated earlier on the biotechnology of history and the nature of science, scientific discovery is a dynamic, laborious, and often tortuous and unpredictable enterprise. The process involves a medley of factors, situations, variables, and many others that remain largely capricious. Undoubtedly, scientific research whether covertly or overtly has paradigmatic imports for human life in general and other lives such as animals as well. As such, the frontiers of biotechnology involve a dynamic process of research and discovery in consonant with general scientific enterprises and principles. Therefore, from basic biotechnology research in laboratories to apply research involving living systems such as microbes and human subjects, there is a general anticipated conduct that scientists ought to follow. Scientific and biomedical researches in particular have reflected some of the basic dictum of the Hippocratic tradition or others. But the code of Hippocrates though seemingly universal does not necessarily have definitive guidelines on conducting scientific and biotechnological research per se. Textually, there is no extant reference and no explicit suggestions on responsible biomedical research. However, one thing seems evident—physicians or healthcare providers' encounter with patients (within the Hippocratic tradition) must reflect and enhance a genuine respect for privacy and above all as Hippocrates noted in *Epidemics, Book I. Section 11: As to diseases, make a habit of two things—to help, or at least to do no harm*, or, as tacitly reflected later in the aphorism, *primum non nocere* (above all *or* first of all do no harm). Furthermore, Aristotle's eponymous work, the *Nicomachean Ethics*, and many other philosophico-ethical works such as those of Kant, Kierkegaard, Spinoza, Thomas Aquinas, Hegel, and Augustine and a cluster of others have formed some of the core ethical compasses in past biomedical research. It is worth noting that by the beginning of the twentieth century, a robust medical professional ethical tradition had already emerged and systemized for physicians as in the AMA's Code of Ethics and the British Medical Association's Code of Ethics, respectively. Basic research in science was guided in a very fragmentary way and to some extent by these and peer reviews, internally regulated laboratory protocols and norms, institutional principles, professional code of conducts, and local norms on hygiene.

This is because research is an intentional act and it must follow certain basic regulatory procedures. Indeed, as the Common Rule (45 CFR 46. 102) noted, *a research means a systematic investigation, including research development, testing and evaluation, designed to develop or contribute to generalizable knowledge. Activities which meet this definition constitute research for purposes of this policy, whether or not they are conducted or supported under a program which is considered research for other purposes. For example, some demonstration and service programs may include research activities.* This is a broad definition of what constitutes research. Biotechnology research ought to reflect these elements of research at any level of its activities. Research is not just an experiment for the sake of curiosity but it must be purposeful. Historically, biomedical research has not consistently reflected this notion of intentionality and implications to contribute to general knowledge. As a matter of fact, prior to this broad definition, researches, especially in biotechnology, were internally regulated from institution to institution, peer reviews, some professional codes of conduct, and general ethical norms that are not necessarily universal and categorically binding to researchers. Therefore, despite all the significant scientific fervor and biotechnological research of the nineteenth to the twentieth centuries, there was a lack of systemized ethical norms to guide research involving human subjects and primates. It is also important to indicate, however, that pharmaceutical research has evolved with a robust "regulatory" tradition since the beginning of the twentieth century until now. While regulatory policies are not ethical per se, nevertheless, they have indices of ethical parameters that many scientific researchers have relied on. Suffice it to say that biotechnological research exhibits unique characteristics. However, several egregious incidents of "research" involving human subjects have led to the promulgation of federal and international norms or codes of conduct to inform and regulate how biotechnological or biomedical research ought to be conducted. Both WW I and II synergize science into a new level of political tool and control. Both the social and political climate encouraged surreptitious research in science. While others invested in huge scientific ventures certain others conducted biomedical researches under the expediency of secrecy to the peril of human life. Shortly after the end of World War II, evidential and compelling news emanated about several surreptitious biomedical researches conducted by Nazi physicians and of scientists on human subjects (mostly prisoners) against their consent in concentration camps territorially under Nazi occupation and control. Myriads of the human research subjects died very horrific deaths. As the world recovered from the rubbles of World War II, it became very clear that these "researches" were actually war crimes under the aegis of science. It glinted renew international efforts and the immediate urgency to bring the perpetrators to justice. As the world pondered over these horrific issues committed in the name of "science," it was evident that regulatory and code of ethics for conducting research were not followed and there was no justification to conduct those researches. As a result, the Nuremberg Trial (1945–1946) court was under the tutelage of the International Military Tribunal with prosecutorial powers to investigate and sought justice for the victims. The *Doctors' Trial* specifically focused on the prosecution and conviction of the physicians and scientists involved in the abuse of human research subjects between 1946 and 1947. A total of 20 physicians and three scientists were tried (*The United States of America v. Karl Brandt* et al.) for inappropriately conducting "experimentations" on human subjects that clearly violated their rights as humans. A 10-point principle formed the ethico-legal basis for the trials and has since been codified as the *Nuremberg Code (NC)* viz:[33]

1. The voluntary consent of the human subject is absolutely essential, etc.
2. The experiment should be such as to yield fruitful results for the good of society, unprocurable by other methods or means of study, and not random and unnecessary in nature.
3. The experiment should be so designed and based on the results of animal experimentation and a knowledge of the natural history of the disease or other problem under study, that the anticipated results will justify the performance of the experiment.
4. The experiment should be so conducted as to avoid all unnecessary physical and mental suffering and injury.
5. No experiment should be conducted, where there is an a priori reason to believe that death or disabling injury will occur; except, perhaps, in those experiments where the experimental physicians also serve as subjects.
6. The degree of risk to be taken should never exceed that determined by the humanitarian importance of the problem to be solved by the experiment.
7. Proper preparations should be made and adequate facilities provided to protect the experimental subject against even remote possibilities of injury, disability, or death.
8. The experiment should be conducted only by scientifically qualified persons. The highest degree of skill and care should be required through all stages of the experiment of those who conduct or engage in the experiment.
9. During the course of the experiment, the human subject should be at liberty to bring the experiment to an end, if he has reached the physical or mental state, where continuation of the experiment seemed to him to be impossible.
10. During the course of the experiment, the scientist in charge must be prepared to terminate the experiment at any stage, if he has probable cause to believe, in the exercise of the good faith, superior skill and careful judgement required of him, that a continuation of the experiment is likely to result in injury, disability, or death to the experimental subject.

The *NC* was the main ethico-legal document used in the trial of the physicians and the scientists for their egregious human experimentations. For the immortal respect for the victims, no specific details of any of the experiments will be recounted here. But there was no doubt about the enormity of the ethical violations of the researches. To curtail these from recurring, the *NC* became a universal ethical code of reference for the conduct for research involving humans. However, the *NC* was not incorporated officially into the official legal lexicons of any single country. Later, the *Geneva Declaration* (1948) attempted to reform the *Hippocratic Oath* and made it more relevant and ethically applicable to contemporary biomedical professionals following the Nuremberg Trials, especially for physicians. Following several revisions to the declaration, it was amalgamated with the ethical principles undergirding the *NC* and the new ethical code became the *Declaration of Helsinki* (*DOH*) (1964). The *DOH* was officially adopted and formed the backbone of ethical conduct that regulates research involving humans globally and the biomedical community in particular. The *DOH* has also undergone many revisions (10 times) though the core mandate has consistently focused on the ethical calculi of balancing good research and protection of human subjects.

As an official bioethical and research document, the *DOH* is thematically structured around the following: risks, burdens and benefits, vulnerable, scientific requirements and research protocols, research Ethics Committees, privacy and confidentiality, informed consent (IC), use of placebo, post-trial provisions, and unproven interventions in clinical practice. As noted earlier, the *DOH* was adapted from the Geneva Declaration, which was initially adopted by the World Medical Association (WMA) in response to the post-Nuremberg Trials. The *DOH* is addressed specifically to guide physicians' ethical and professional conducts in biomedical research. As the preamble to the *DOH* in pertinent parts notes: *The World Medical Association (WMA) has developed the Declaration of Helsinki as a statement of ethical principles for medical research involving human subjects, including research on identifiable human material and data* (Article #1). The *DOH* thus covers a broad spectrum of biomedical research. Rather than a de facto document, the *DOH* is generally considered a "statement of ethical principles." As an ethical principle, it gives both specificities and allows for adaptability of the statements to suit local and international ethical standards and situations. The *principles* referenced in the preamble are quite broad and reflects generally acceptable ethical theories and praxis that were conspicuously violated in the pre-Nuremberg era.

In addition, the *DOH* enunciates and expatiates on these principles. First, risks, burdens, and benefits of biomedical research involving human subjects. The *DOH* notes that every biomedical research involving human subjects have some modicum of risks as such, researchers must determine the burden and benefits in view of potential risks prior to the research. As a result, if there are alternative treatments that confer lesser risks and confer perhaps the same therapeutic benefits, it will be unethical to continue with further research. This is clearly noted in article 7 of the *DOH*: *All medical research involving human subjects must be preceded by careful assessment of predictable risks and burdens to the individuals and groups involved in the research in comparison with foreseeable benefits to them and to other individuals or groups affected by the condition under investigation.*

Second, the *DOH* also discusses research involving "vulnerable" subjects in society. Generally, vulnerable groups include children, prisoners, the aged, women (in some cultures), and others. There are only two reasons for conducting research in a "vulnerable" population. The group/population must medically *benefit* from the research and the knowledge must be of significance to them. As such, researchers must explore human subjects first from nonvulnerable groups prior to considering the vulnerable for research. This is to insulate vulnerable population from potential risks and abuse as happened with the infamous thrombospondins (TSP) and the *Porton Down Chemical Experiments* (PDC) in the United Kingdom, which spanned from 1939 to 1989. In the PDC, over 10,000 military men were treated with the deadly mustard and nerve gases. By all standards, the researches knew the irreparable dangers in the PDC experiment nonetheless; they carried it on "military men" who typically have an unalloyed allegiance to their superiors. Another recent research conducted on "vulnerable population" was reported in the June 2014 edition of *The Lancet*. The researchers recruited 2000 children in India to test an experimental drug for treating *rotavirus* (generally regarded as life-threatening virus infection). It should be noted that rotavirus vaccine and other effective therapeutic interventions were already in existence. However, researchers randomized and placebo controlled the research, which means some of the research subjects were not given the previously known treatment during the trials thus exposing them to extenuating danger in tacit violation of the *DOH* article 7. Surprisingly, the later experiment was approved by an Institutional Review Board (IRB)!

Also, the *DOH* stipulates that a committee must approve research protocols. Typically, an IRB will be responsible either in the originating country or at the international location or both depending on the level of collaboration between researchers and local folks. The *DOH* also stipulates proper scientific justification for the research, the use of placebo, IC, respect, privacy, and confidentiality of research subjects. Furthermore, researchers must make trial provisions *in advance of a clinical trial, sponsors, researchers and host country governments should make provisions for post-trial access for all participants who still need an intervention identified as beneficial in the trial. This information must also be disclosed to participants during the informed consent process* (*DOH* #34).

Besides, biotechnological innovations and research was undoubtedly rapid in the 1970s, as noted in my earlier reflections. This rapidity was rift especially in the emergence of rDNA biotechnology. The science and skills for cloning, species, and interspecies hybridizations were no longer scientific fictions but existential realities. Both within the scientific community and the legislatures, concerns were being raised about the direction of biotechnological/biomedical research. In California, scientists had gathered for the *Asilomar Conference on Recombinant DNA* on these new and unique challenges inherent in biotechnology. Some of the issues discussed were on risks and biosafety especially in cloning potentially biotoxins and rDNA-based experiments that posed risks to humans. The general conclusion was to adopt a precautionary approach in the conduct of biotechnological researches. In addition, reports emerged about an equally egregious human research conducted on poor black males, the *Tuskegee Syphilis Project* in the 1970s.[34] The original research spanned from 1932 to 1972. Words can never describe the callousness of the TSP because, the cure for syphilis was already known, and the NCs as well as the Geneva Declaration and *DOH* were already public in the public domain and internationally recognized as bioethical documents to protect the human subjects. These revelations and the fact that there were some survivors led to a federal investigation. The *National Commission for the Protection of Human Subjects of Biomedical and Behavioral Research* was constituted by the federal government to develop a code of research ethics to address these abuses and to further protect future research involving human subjects. The commission was mandated to consider:

1. The boundaries between biomedical and behavioral research and the accepted and routine practice of medicine
2. The role of assessment of risk–benefit criteria in the determination of the appropriateness of research involving human subjects
3. Appropriate guidelines for the selection of human subjects for participation in such research and
4. The nature and definition of informed consent in various research settings.

The deliberation took place in Belmont, Maryland, and led to the eponymous *Belmont Report,* which the commission presented in 1979. Among other things, the report highlighted the philosophico-ethical principles for the autonomy of humans, beneficence and nonmaleficence and justice. These constitute what has become "principlism." Practically, these principles translate into IC, assessing risks and benefits and the selection of research subjects involving humans. The *Belmont Report* has been promulgated as the *Code of Federal Regulations* (45 CFR 46) by *Department of Health and Human Services* (HHS). Biotechnological researchers are bound to adhere by these bioethical norms and codes in the United States. The norms are routinely embodied into other international ethical codes for the protection of human subjects and ensuring the integrity of research in particular.

End notes

1. Sandra Shapshay. *Bioethics at the Movies* (The Johns Hopkins University Press; Baltimore, MD, 2009), pp 75–78.
2. J. Langheier et al. Prospective medicine: The role for genomics in personalized health planning, *Pharmacogenomics* 5/1: 2004, 1–8. Lehninger, *Principles of Biochemistry* (W.H. Freeman and Company; New York, 2008), pp 275–299.
3. Susan E. Lederer. *Subjected to Science: Human Experimentation in America before the Second World War* (The Johns Hopkins University Press; Baltimore, MD, 1997), p 1.

4. Bonnie Steinbeck, *The Oxford Handbook of Bioethics* (Oxford University Press; London, 2007), p 472.
5. Ibid.
6. Ibid.
7. Dennis K. Flaherty et al. Single nucleotide polymorphisms, drug metabolism and untoward health effects, *Journal of Medical and Biological Sciences* 1(2): 2007, 1–8.
8. Ann K. Daly et al. CYP2C9 polymorphism and warfarin dose requirements, *Journal of Clinical Pharmacology* 53(4): April 2002, 408–409; Michael D. Caldwell et al. Evaluation of genetic factors for warfarin dose prediction, *Clinical Medical Research* 5(1): March 2007, 8–16; Ann K. Wittkowsky. *Warfarin* (AHFS 20:12.04).
9. Ibid.
10. Ibid. See also Francis Collins et al. Genomic medicine—An updated primer, *New England Journal of Medicine* 362(21): May 27, 2010, 2001-2011.
11. Adrian Dubock. The politics of Golden Rice, *Journal of GM Crops & Food* 5(3): 2014, 210–222.
12. Francis Collins. Medical and societal consequences of the human genome project, *New England Journal of Medicine* 341: 1999, 28–37.
13. www.genome.gov
14. Ibid. Arthur L. Caplan et al. Ethical considerations in synthesizing a minimal genome, *Science* 10: December 1999, 2087–2090; Jeremy Sugarman. Ethical considerations in leaping from bench to bedside, *Science* September: 1999, 2071–2072; James M. Jeffords et al. Political issues in the genome era, *Science* February: 2001, 1249–1251; Susan M. Wolf. Patient autonomy and incidental findings in clinical genomics, *Science* May: 2013, 1049–1050: Francis Collins. Medical and societal consequences of the Human Genome Project, *New England Journal of Medicine* 341: 1999, 28–37; Susan E. Lederer. *Subjected to Science: Human Experimentation in America before the Second World War* (The Johns Hopkins University Press; Baltimore, MD, 1997), p 1.
15. http://ghr.nlm.nih.gov/handbook/hgp/elsi (The *Human Genome Project: Ethical and Social Implications.*) See also Thomas A. Shannon et al. *An Introduction to Bioethics* (Paulist Press; New York, 2009), pp 107–250.
16. Dubock. *The Politics of Golden Rice.*
17. Robert Bud. *The Uses of Life: A History of Biotechnology* (Cambridge University Press; London, 1993) and Robert Bud. *History of Biotechnology* (John Wiley & Sons Ltd; Chichester, 2003).
18. Ibid
19. J. Huxley. *Retreat from Reason*, ed. Hogben L.I., (Random House; New York, 1938).
20. Bud. *History of Biotechnology.*
21. Bud. *History of Biotechnology.* See also A. Fiechter. *History of Modern Biotechnology* (Springer; New York, 2000).
22. Bud. *History of Biotechnology.*
23. James Pritchard. *Ancient Near East. An Anthology of Texts and Pictures* (Princeton University Press; Princeton, NJ, 2011); Marc Van De Mieroop. *A History of Ancient Near East: ca 300–320 BC* (Wiley Blackwell; Chichester, 2016).

24. Albert Jonsen. *The Birth of Bioethics* (Oxford University Press; New York, 1998).
25. www.bisworld.com/industry/report/global/biotechnology
26. Lisa A. Borgerding et al. Teachers' concerns about biotechnology education, *Journal Science Education Technol*ogy 22(2): 2013, 133–147.
27. Alan Michette and Sławka Pfauntsch. *X-Rays: The First Hundred Years* (John Wiley & Sons; Chichester, 1996).
28. Thomas Kuhn. *The Structure of Scientific Revolution* (University of Chicago Press; Chicago, IL, 2012), p 69.
29. Bud. *History of Biotechnology.*
30. Kuhn. *The Structure of Scientific Revolution.* See also, William F. McComas et al. *The Nature of Science in Science Education: Rationales and Strategies* (Springer; Dordrecht, Netherlands, 2006), p 511.
31. Van Rensselaer Potter. *Bioethics: Bridge to the Future* (Prentice-Hall Press; Englewood Cliffs, NJ, 1971) and V.R. Potter. Bioethics, science of survival, *Perspectives in Biology and Medicine* 14: 1970, 153.
32. Onora O'Neil. *Autonomy & Trust in Bioethics* (Cambridge University Press; London, 2002), p 1.
33. George J. Annas and Michael A. Grodin. *The Nazi Doctors and the Nuremberg Code: Human Rights in Human Experimentation* (Oxford University Press; New York, 1992).
34. Susan M. Reverby. *Tuskegee's Truths: Rethinking the Tuskegee Syphilis Study* (University of North Carolina Press; Chapel Hill, NC, 2000).

SECTION II
Bioethics and Genomics

2
Biotechnology and Bioethics

Introductory comment

Advancements in molecular genetics coterminous with the computational capacity to process high throughput biodata have given new impetus for charting a new paradigm in pharmacogenomics and PM. First, the ability to study and sequence genetic materials explicates the intricate complexity and organization of life at the molecular level. Through the completion of the sequencing of the entire human genome, scientists are able to extrapolate and analyze genotypic data, genetic variants such as *single-nucleotide polymorphisms* (SNPs), genomic mutations and deletions, and identify the potentials these hold for the advancement of biomedical science. Second, gene modification tools especially restriction enzymes used in recombinant biotechnology has been improved and are available for application in clinical, research, and in the development of the next generations of biologics. Third, as a matter of extraordinary coincidence, computational technologies such as *new generation sequencing* (NGS) also advanced concurrently cascading in the ability to develop *in silico* models to augment clinical research in drug discovery and also to process, store, and analyze the myriads of data generated through genomic study thus expanding the scope of bioinformatics and synthetic biology. This epistemic shift has also resulted in a better understanding of the functions of genes and the molecular pathways undergirding the etiology of atypical diseases and the potentials for tailoring specific therapeutic and pharmacological interventions. These have also generated concatenations of ethical trepidations and policy concerns. How would genomic information be used and by whom? What are the justifications for gene editing such as embryonic cells? Are there legal frameworks and traditions in ensuring the protection of vulnerable populations given historic precedents such as eugenics? Does an individual have the autonomy and legal authority to make decisions on his genomic data? How accurate are these genetic tests and how could these be regulated? These and a medley of other ethical conundrums would constitute the foci of this chapter.

The human genome comprises approximately 20,000 coding genes with three billion DNA basepairs.[1] It is estimated that only 1%–1.4% of the human DNA encodes for proteins or serve any known useful purpose. However, within the human genome, there are DNA sequence variations known as SNPs.[2] SNPs occur every 300 nucleotides hence about 10 million of these exist in the entire human genome within the coding and noncoding regions of genes.[3] SNPs are loci in the gene in which a basepair is different in individuals within a population. They do not have any effect on cellular functions per se. For example, a DNA sequence might have read AGTT**C**GATGCG for a particular protein but in another person, it might be AGTTAGATGCC.[4] In this example, the nucleotide cytosine in the fifth position had changed to an adenine! This slight aberration accounts for some of the genetic variations within the human species, manifesting in phenotypic indices such as differences in hair color, foot sizes, allergies, and reactions to medications.[5] SNPs are conserved in the evolutionary trail hence it's importance as a biomarker in forensics, pharmacogenomics research and applications in PM. In addition, genes may undergo spontaneous mutations due to many factors and these could potentially result in serious neurogenerative clinical conditions as well as cancers. Gene editing tools could repair these mutated genes. All of these are giving new therapeutic prospects in precision medicine, precision nutrition, pharmacogenomics, neuroscience, epidemiologist, and public health among others.

Pharmacogenetics may be defined as the "...study of differences among a number of individuals with regard to clinical response to a particular drug," and *pharmacogenomics* comprise the "...differences among a number of compounds with regard to gene expression response in a single (normative) genome."[6] As some scholars have adroitly indicated, both pharmacogenomics and pharmacogenetics "attempt to elucidate the role of genetic variation in human responses to compounds introduced into the body, such as medications."[7] In this book, I will use both words interchangeably without prejudice. Another concept worth defining is PM. Generally, PM entails tailoring therapeutic interventions to individuals or defined populations because of specific biomarker. According to the National Research Council, PM is the use of "genomic, epigenomic exposure and other data to define individual patterns of diseases, potentially leading to better individual treatment."[8] Advancements in the study of molecular genetics have thus ushered in an unprecedented era that will continue to define the way medicine is practiced. In brief, the symbiotic relationship between pharmacogenomics and PM has generated some ethical, regulatory, and legal debates and I intend to focus on these respectively.

Autonomy

One of the stalwarts of computational engineering, Steve Jobs' death sparked a flurry of ethical debates on the issue of patients' rights and capacity to make autonomous decisions. He was diagnosed with a rare form of pancreatic cancer (islet cell neuroendocrine tumor) in 2003. Against his physicians' advice, Jobs made an autonomous decision and refused the standard treatment for cancer for several months. Even when his colleague, Tim Cook offered part of his liver to him for transplant after his initial surgery, he still made the "decision" and declined the offer. It has been generally believed these decisions caused him his life. Could Jobs have saved his life if he had accepted the physician's advice? Could the physicians have "compelled" him to undergo the initial therapeutic procedures?

As with many medical decisions, physicians are obligated to respect the autonomous decision capacity of their competent patients irrespective of the consequences. In this chapter, I examine the notion of autonomy and its paradigmatic import to genomic and PM, in particular.

First, let us examine the ethical principle of *autonomy*. The concept of autonomy has been popularized in contemporary times. People assert and make claims about all or some aspects or spheres of their lives that they want protected unreservedly devoid of any interference by any entity. People want to make decisions about their lives. Such decisions could be about their own medical, social, and educational situations to the extent that they are free to do so without someone presumably competent in that field making that decision for them. While making autonomous decision does not necessarily guarantee that the outcome of such decisions may be desirable, nonetheless, the capacity and the ability to do so is deeply aligned with the individual's liberty and right as free and existential beings. Autonomy has become one of the most significant principles in the corpus and genre of bioethics in contemporary times. This principle was carefully dissected during the debate leading up to the *Belmont Report* generally assumed to have laid the foundation stone for modern bioethics. Furthermore, there is an insatiable patient population that is increasingly well educated or informed about biomedicine and capable of challenging medical professionals about decisions pertaining to them. This may be partly accentuated by the fact that medical paternalism is increasingly becoming a residue or a footnote in biomedical practice with patients'/research subjects asserting their rights. The epistemic shift in medical paternalism to respecting and according patients their autonomy has a wide scope of implications for genomic medicine and clinical research, in particular. But what then is autonomy? What are the philosophical underpinnings of the concept? Etymologically, the word autonomy comes from two Greek words: *autos* (self, individual) and *nomos* (norms, rule, and laws). By juxtaposing these two words, *autonomy* literally entails *self-regulation or self-governing*.[9] That is the capacity and the ability of an individual to make his own norms devoid of external interference in as far as it does not harm any other person. Competent individuals make many decisions each day; what to wear, what they want to eat, who they want to associate with, what kinds of profession that want, and so on under the assumption that they are free to do (deservedly) so. Autonomy is also premised on the notion that every human being has an intrinsic capacity to make decisions that insure their welfare, safety and happiness devoid of undesirable restraints. Within the ethical corpus of "Principlism," autonomy is the idea that every human being has the inherent capacity to make informed decisions and self-regulate as applicable to medicine and other areas of his life. This is significant because every individual is a complete and whole being whose rights, choices, decisions, and actions are intrinsically imputed to him. That is to say, every human being has complete authority to make informed decisions. Such decisions may oscillate on the ability to withdraw from a medical procedure, clinical research, or the decision to opt out for alternative medical intervention with the case of Steve Jobs who opted for pseudo medical treatments upon his initial diagnosis.

Second, the concept of "autonomy" may be traceable to ancient Greeks especially in the writings of Plato and Aristotle.[10] Autonomy constitutes self-mastery or the ability of the rational part of the individual to subjugate the nonrational part. Autonomy implies an individual or the small Greek cities do not rely or depend on others for self-determination especially in the context of ethics in order to attain happiness (eudemonia).[11] This idea became a central theme in the corpus of Kantian ethics. Kant challenged the hitherto idea of individual's reliance on external authorities/sources in making ethical decisions.[12] According to Kant, human beings depend on *posteriori* maxims or external principles in making rational decisions.[13] When these subjective maxims become universal, that is if everyone, anywhere in the world would abide by such maxim(s), then they have attained a universal moral principle or a categorical imperative that binds all people. In other words, *autonomous* beings are rational, self-legislative beings that make their *own decisions* devoid of external interference. This concept of individual autonomy has transcended many ethical schools of thought. However, Mill challenged the Kantian concept of "individual autonomy" and emphasized the significance of both rational and nonrational part of the person: "A person whose desires and impulses are his own—are the expression of his own nature, as it has been developed and modified by his own culture—is said to have a character. One whose desires and impulses are not his own has no character, no more than a steam engine has a character."[14] The emergence of "biomedical ethics" has galvanized the concept of "autonomy" and changed the scope of the discourse.[15] In contemporary times, several strands of the concept of autonomy have emerged.[16] Indeed, Beauchamp postulated autonomy as both positive and negative within the context of *Principlism*.[17] In a positive sense, physicians and researchers are required to respect decisions made by their patients or research subjects. In a negative sense, autonomy entails absolute protection from external or any interference in biomedical decision process. In brief, autonomy means self-legislating or self-governance.

These analyses have several implications for genomic and the evolving nature of PM. The concept of autonomy is increasingly used to justify why individuals must give their consent in biomedical research and therapies. Every individual has absolute control over his genetic data or information and other bodily specimens extrapolated from him for either research or therapy. The decision to make these data available to the general public or researcher resides in the autonomous individual. But this was challenged during the landmark case of *Moore v. Regent of University of California, Los Angeles.*[18] Moore contested in court that his autonomy to give IC was violated because his attending physician did not disclose to him that his biospecimen were being used for genetic clinical research worthy of any commercial value. Even though the court ruled against him, many bioethicists and scholars have challenged the tacit violation of his autonomy and his ability to self-determine what his biospecimen and genetic data could be used for.

Furthermore, one of the challenges to the concept of autonomy is what some scholars call geneticization.[19] As one scholar, Henk ten Have, has noted, "Geneticization is the sociocultural process of interpreting and explaining human beings using the terminology and concepts of genetics, so that not only health and disease but all human behavior and social interactions are viewed through the prism of biomolecular technology."[20] Although benign or unintended, genomic medicine seems to *objectify* an individual to be mere constituents of genes. If genes are *a priori* entities (existing prior to the individual's ability to use reason), it would seem subtle to suggest that autonomy might not actually truly exist because individuals are merely constituents of their genes and cannot fully make their own autonomous decisions since they are predetermined by genetic proclivities. If these assertions are true, then strictly speaking, no individuals can be truly construed to be genetically autonomous and cannot therefore make their own decisions especially in the context of genomic and precision medicine. While to some extent this might be true, genetics does not always and fully define the autonomous person. An individual is a combination of his gene and interaction with his environment. Every human being then can and does make decisions; hence, genetics alone cannot define the individual person. In genomic medicine, individuals would still have to make decisions about what kinds of genetic tests they want to undergo and whether they want to give consent on what kinds of their own genomic information they want to share. Respect for individual autonomy should be in the broader context of society. This is because the society in which the individual was born and nurtured has social values that shaped that individual and is significant in addition to his genetic profiles.

In addition, the question of autonomy becomes even complicated in the contexts of *Genomic Wide Association Study* (GWAS) and precision medicine.[21] This is partly due to the perception that genetic traits generally are not only the preserve of individuals per se but also are shared entities within a specific population. People of specific genotypes have some common genetic traits and sometimes these become critical in developing specific pharmacogenomic interventions. For example, *sickle cell gene* (HbS gene) is estimated to be present in over 25% of the entire population in West Africa and endemic in other parts of the world.[22] Because of the prevalence of sickle cell in this population, no individual can make a claim of autonomy per se when there are genome-wide studies of this disease. This then leads to the claims that individual autonomy is invariably intertwined with societal autonomies. Hence, respect for individual autonomy should be coterminous with respect for "social autonomy." After all, as the aphorism goes, *no one is an island* since individuals share significant genetic information with each other. Consequently, equal and dignified consideration should be given to society or subgroups of people who might share some genetic traits. Should social autonomy supersede individual autonomy? To what extent should this be especially in the context of genomics medicine? What are the bounds and mints of autonomy? Does the individual have the fiduciary obligation to inform society of potential adverse genetic information? The concept of social autonomy imposes some ethical and fiduciary obligation on an individual to inform society or the public about some genetic information that is clinically actionable. This is because to some extent, society protects and respects individual autonomy. In addition, because genetic is a shared entity, members of society have some natural and proximal obligation toward each other and this must be considered as well. For example, Angelina Jolie, the celebrated actress decided to test for mutations in the BRCA 1 gene responsible for some breast cancers because some of her relations had breast cancer. But because they disclosed their individual medical information with her, she acted promptly for the test which turned out positive. She was eventually treated. If the family members had kept it "secret," she probably might not have known. The contention of this piece is that respect for individual autonomy is invariably linked to respect for social autonomy especially in the context of genomic medicine. Each exists independently but depends on each other. In a society that triumphs respect for *individual autonomy,* this will continue to pose challenges especially in the development of precision medicine.

In brief, autonomy in a *sui generis* sense is significant especially in genomic medicine because individual and social autonomies cannot and do not exist in isolation. An attempt to isolate these could and would continue to pose ethical tensions. In biomedical research, we see these challenges coming to the fore when individual consent and social values become dialectically opposed to each other. Should individual consent take precedent over social or societal consent? Why and how? What about disclosures of genetic information? Responses to these would constitute the topic for our next discussion.

Confidentiality and privacy

In a recent breach of medical confidentiality, former Massachusetts Gov. Wilson Weld's medical records were extrapolated from purportedly de-identifiable information on the web and made publicly available by Latanya Sweeney.[23] As Sweeney later testified, nearly 4 in 10,000 patients' confidential medical records are clandestinely extrapolated from the web with ease and alacrity. As a matter of fact, "Several large, health-related data breaches received attention from the Department of Justice. In one case, 6,800 electronic health records of patients at New York-Presbyterian Hospital and Columbia University ended up on Google due to inadequate safeguards, resulting in a $4.8 million fine. And a $1.7 million fine was levied against the health insurance company WellPoint, after health data, along with social security numbers and demographic information, were made accessible on the internet, where it resided for almost five months."[24] Obviously, this raised public ire and furor because healthcare records are presumptively protected by a consistent tradition of confidentiality and privacy. In an era of genomic medicine especially big biodata and the advancement of computer technology, this poses some ethical strains. These challenges become even compounded during pharmacogenomics research. Confidentiality is a cherished value and a bastion of the researcher–subject relationship (RSR).[25] The concept of confidentiality is evolving and multifaceted with a wide scope of applications especially in the burgeoning field of pharmacogenomics and PM. It has metamorphosed with the emergence and development of medicine since ancient times and throughout many cultures. Sometimes, confidentiality is synonymously associated with secrecy and privacy even though they generally are not. But the concept has gained much usage and I believe it is important to clarify these.

Confidentiality and privacy have been significant bedrocks of patient–physician relationship (PPR) in the Hippocratic medical tradition. As Hippocrates once said, "Whatever, in connection with my professional practice or not in connection with it, I see or hear in the life of men, which ought not to be spoken of abroad, I will not divulge, as reckoning that all such should be kept secret."[26] The AMA renders the same quote as "That whatsoever you shall see or hear of the lives of men or women which is not fitting to be spoken, you will inviolably keep secret" (AMA).[27] There is thus a subtle professional injunction and obligation for physicians and healthcare providers to keep information that are of both medical and nonmedical domains about their patients confidential. The PPR is solidified under the expediency of trust and mutual respect that information regarding a patient's medical condition and other spheres of his life are protected from a third party or as so defined within an agreeable context. Confidentiality ensures that patients are able to open up and be truthful in disclosing medical information to their physicians for appropriate therapeutic intervention and management. It is important to point out here that while not absolute, confidentiality and privacy within PPR fosters a modicum of reciprocity because the patient ought to keep confidential medical issues to the extent allowed by law. But the concept of confidentiality has a wide scope of meaning and applications depending on many parameters.

The notion of confidentiality has been clearly documented and appropriated in every epoch of medical advancement and history.[28] That is to say, despite the enormous stride in the development of medicine, confidentiality within the PPR has accentuated every aspect of biomedical profession till this day. The significance of PPR is further bolstered in *The American Medical Association's Code of Medical Ethics* (AMACME) when it states … "Information disclosed to the physician during the course of the relationship between the physician and patient is confidential to the greatest possible degree." That is to say, there is a binding professional code of conduct for physicians. Akin to the Hippocratic Code, the AMACME anticipated PPR will generate some modicum of information that must be protected at all times especially by physicians because, as Kottow noted, "Clinicians' work depends on sincere and complete disclosures from their patients; they honor this candidness by confidentially safeguarding the information received."[29] Furthermore, protecting PPR insulates patients from any perceived or existential harm. Hence, breaching PPR has been consistently considered a tacit violation and professional misconduct with preponderant consequences. In addition, the Geneva Declaration imposes a confidential obligation on physicians to protect patients' clinical information throughout their lifespan and even posthumously as encapsulated in these words, *I will respect the secrets confided in me, even after the patient has died.*[30]

Confidentiality within the corpus of *contractual law* implies that physicians have a *prima facie* obligation to their patients especially their medical information from getting into the hands of a third party without their expressed, verbal, written, or proxy consent. This is because the physician–patient relationship is a choice. A patient chooses his own healthcare provider freely without duress. The healthcare provider does not have any obligation to accept him (of course except in the context of an emergency). Because such choices are presumptively free choices by competent people or by their proxies, a *sine qua non* contractual obligation then begin to exist between them. Among these are the issues of confidentiality, prevention of harm, and the obligation of beneficence.

In fiduciary theory also, healthcare providers have the obligation to protect the most vulnerable in their care and herein the patients. In common law as well as federal/state laws, healthcare providers are required to protect the privacy and confidentiality of their patients at all times (except where there are incidence of infectious diseases and epidemics). For instance, The Health Insurance Portability and Accountability Act of 1996 (HIPAA) was enacted amidst concerns about the potential breaches of patient's private medical records. Many states also have formulated their laws to align with HIPAA to ensure even greater protection of confidential medical records of patients. This is critical especially in an era fraught with information overload. According to a recent *Pew Research* survey, patients' perception and concerns about medical confidentiality seem to vary. About 20% of the US adult respondents worry about their clinical confidential information being with their healthcare providers, while 27% do not worry that much, nearly 30% have confidence that their confidential medical records are safe.[31] In a related Gallup poll, 93% of US adults felt that researchers, healthcare providers, and governments should first seek permission from patients prior to accessing their genetic confidential records or information.[32] And 82% of patients object to insurance companies having access to their medical records devoid of their permission. In brief, while patient's views, concerns, and perceptions vary on the central issue of medical confidentiality and privacy, the surveys nonetheless affirm the fact that in any measure, they want to have active and full participation on the use and disclosure of these information other than healthcare providers.

In response to these questions and challenges and varying perceptions, some scholars have called for absolute confidentiality (AC) while others argue for some kind of middle ground. First, let us examine the question of *AC* in genetics test and medicine. Proponents are of the view that genetically identifiable medical information of patients should be protected all the time from third parties even though relatives may have similar genetic traits that may be good or medically actionable. Such arguments have strong theoretical, ethical, policy and legal underpinnings of respecting individual patient's autonomy rather than social autonomy. An individual's genomic and clinical information (though significant to his/her extended family members) should not be divulged unless expressly required by a statutory norm or law. In addition, AC may also imply *indefinite confidentiality.* In other words, confidentiality within the scope of PPR and the RSR should not transcend all time and may not even be disclosed posthumously. Therefore, in case there is a potentially debilitating genetic information, family members may be precluded from having access even posthumously. It is important to point out here that some moderates of this view argue for confidentiality within specific scope and framework that allows for disclosures but not indefinite. That is upon the death of a patient or with due diligence in obtaining consent, genetic information may so be disclosed to any competent and legally responsible person or entity provided such information are protected at all times. In brief, the question of disclosure of genetic information is seemingly redefining the scope of medical confidentiality within the contexts of physician–patient relationship. In brief, two strands and schools of thoughts emerged: absolute but not indefinite confidentiality and indefinite confidentiality in the contexts of genetic medical information.

Another approach inferred from the above discussion is that some moderates argue for *contextual confidentiality*. That is to say, there is a presumptive confidentiality in the PPR at all times since this is critical in the overall healthcare delivery. A patient must truthfully and sincerely disclose medical information; either good or adverse information including those of his family to his physician for proper diagnosis or prognosis with a view to improving his overall health index and of society, in general. Furthermore, *contextual confidentiality* implies that under certain contexts or circumstances as permitted by law and professionally justified, physicians or healthcare providers may disclose such medical information to third parties or people who might be in imminent danger or are vulnerable. Generally, the rationale for the disclosure or breach of confidential medical information include the following: protecting innocent people, public or institutional interest, where the patient might be in danger—paternalistic breaching, posthumous disclosures (although forbidden by the *Declaration of Geneva*), conflicting and competing interest especially in the contexts of epidemics or lingering highly infectious diseases or genetic traits such as the BRCA 1& 2 genes. There is no doubt that the protection of all people including patients and research subjects are symbiotically essential in our society. Confidentiality under the auspices of PPR and RSR does not extirpate this social responsibility imposed on healthcare providers or allied healthcare practitioners or biomedical researchers. Indeed *US Code 38 §7332* entitled *Confidentiality of Certain Medical Records* in pertinent parts states,

> (a) (1) Records of the identity, diagnosis, prognosis, or treatment of any patient or subject which are maintained in connection with the performance of any program or activity (including education, training, treatment, rehabilitation, or research) relating to drug abuse, alcoholism or alcohol abuse, infection with the human immunodeficiency virus, or sickle cell anemia which is carried out by or for the Department under this title shall, except as provided in subsections (e) and (f), be confidential, and (section 5701 of this title to the contrary notwithstanding) such records may be disclosed.[33]

The physician or healthcare provider has a duty to breach PPR confidentiality in order to disclose critical genetic medical information to third parties or competent authorities if there is a presumption that the contrary will pose danger to innocent people. For example, in *Pate v. Threlkel*, the court asserts that the physician has the "duty" to warn both patients and relatives of potential transferable genetic traits or disease. Marriane New was diagnosed with *medullary thyroid cancer* (MTC) caused by mutations in the rearranged during transfection (RET)-proto oncogene (even though, there are sporadic MTC). Shortly after her treatment, her daughter Heidi Pati sued Marriane's physician alleging that she should have been informed since MTC was a genetically transferable disease. The court determined *inter alia* that the physician had the duty to warn both patients and blood relations of the genetic conditions or third parties. Furthermore, the court indicated that the duty to warn was "obviously developed for the benefit of the patient's children as well as the patient"! But the court also indicated that a physician's duty to the patient's blood relations is satisfied by warning the patient of the nature of the disease. It is the onerous responsibility then of the patient to convey the rest of the confidential genetic information to his family as expediently as possible if there is need for imminent clinical actionability.

Critiques of the ruling and the law have been swift in pointing out that such genomic medical information unlike others is unique and different. Some genetic medical information may be actionable or not actionable depending on many factors. For example, a patient suffering from Huntington's disease (HD) may not necessarily be cured by any known therapeutic means. At best, the HD could be managed based on current scientific scholarship and clinical practice. Divulging such confidential genetic information to family members could potentially be a source of stressor such as anxiety, depression, and psychosomatic problems. Hence, genetic information should have unique confidentiality criteria for disclosures to third parties including familial relations. As Offit and others have noted, "In general, the special nature of genetic tests has been viewed as a barrier to physicians' breaching the confidentiality of personal genetic information. However, the failure to warn family members about hereditary disease risks has already resulted in three lawsuits against physicians in the United States."[34] The AMA recognized this uniqueness and challenges and consequently issued a guideline titled, *Disclosure of Familial Risk in Genetic Testing*. The first article of the guidelines states that "Physicians have a professional duty to protect the confidentiality of their patients' information, including genetic information."[35] It clearly reaffirmed the absolute fiduciary obligation of the physicians in PPR confidentiality in contradistinction to the injunction imposed by the *Pate v. Threlkel* rulings. However, the same guideline also states that

> Physicians also should identify circumstances under which they would expect patients to notify biological relatives of the availability of information related to risk of disease. In this regard, physicians should make themselves available to assist patients in communicating with relatives to discuss opportunities for counseling and testing, as appropriate.[36]

In other words, physicians have a dual responsibility toward protecting the confidentiality of their patients and familial relations in medical or health-care issues equivocating on genetics.[37] As a matter of fact and by virtue of professional prudence they are to encourage their patients for the disclosure of potentially adverse genetic test results and information to their relations and as Offit and scholars such as Sam Turner and Eve A. Wadsworth have noted, patients should not be coerced to breach or disclose genetic information.[38] They should do so out of their own volition and if possible in consonant with their physicians. In addition, they are enjoined to avail themselves to discuss genetic test results with their families who might be at risk of potentially inheritable and transmittable diseases. Indeed the veil of the traditional notion of AC has somehow been denigrated in current clinical practice to the extent that physicians could justifiably initiate some disclosures of confidential genetic information with their patients and familial relations. This has been affirmed by The American Society of Human Genetics Social Issues Subcommittee on Familial Disclosure, which also issued its detailed and more comprehensive professional guidelines titled, *Professional Disclosure of Familial Genetic Information,* which in pertinent parts notes as follows:

> Genetic information, like all medical information, should be protected by the legal and ethical principle of confidentiality. As a general rule, confidentiality should be respected. In the context of medical information, privacy rights translate into protection of personal data, affirmation of confidentiality, and freedom of choice. However, the principle of confidentiality is not absolute, and, in exceptional cases, ethical, legal, and statutory obligations may permit health-care professionals to disclose otherwise confidential information.[39]

Furthermore, the documents also proposed some parameters and circumstances under which healthcare providers and researchers may disclose genetic information. For example, it encourages physicians to disclose if patients have failed to inform at-risk relatives of genetic medical information such as those that are imminent, foreseeable, the disease is preventable, and most importantly, the relatives are identifiable and reasonably reachable in terms of reasonable physical proximity.[40] That is, the initiative to breach and disclose confidential medical documents should emanate from the health-care providers rather than the patients (especially if the patient refuses to do so or is reluctant)! Indeed, these and other professional guidelines have been recommended earlier by The President's Commission for the Study of Ethical Problems in Medicine and Biomedical and Behavioral Research. The commission recommended that healthcare professional disclosure to at-risk family members should take place only when (1) *reasonable efforts to elicit voluntary consent to disclosure have failed;* and (4) *appropriate precautions are taken to ensure that only the genetic information needed for diagnosis and/or treatment of the disease in question is disclosed.* In brief, advancements in genetics, pharmacogenetics have brought about some kind of axiomatic shift in health care to the extent that questions of confidentiality and privacy have been subjected to extensive and renewed ethical and medico policy and legal debates. These views and discussions have helped shape these professional guidelines discussed above. We can synthesize the following as plausible criteria in forming and informing decisions about confidential disclosure of genetic information:[41]

1. Probability of harm
2. Magnitude of harm
3. Foreseeability of harm
4. Preventability of harm
5. Identifiability of victim(s)
6. Potential impact on a general policy of confidentiality

These criteria become relevant if the genetic information is medically actionable or not. The question then arises as to whether genetic medical information about BRCA 1 gene and Turner syndrome should be disclosed in the same manner in view of these guidelines. What about *iatrogenesis* or diseases caused due to mutagenesis induced by exposure to environmental carcinogens such as UV lights or heavy chemicals? In response to these, I propose some kind of common ground under the expediency of *genomic clinical pragmatism* (GCP) in the disclosure of genomic information by healthcare providers. Two significant questions herein emerge: are the confidential genetic information to be disclosed important, clinically valid, and actionable? Are the confidential genetic information important but not actionable?

Let us look at what constitutes *actionability* and how this could define the quiddity of disclosing confidential genetic information to third parties. Actionability implies that in a clinical genetic test, diagnosis need immediate therapeutic intervention by a competent medical or healthcare practitioner such as a physician. Furthermore, the risks for the disclosure should be undisputable and clinically validated and if possible a second opinion sought in order to eviscerate any iota of doubt. These criteria should also be applicable during pharmacogenomics research where genetic information are obtained from research subjects for research purposes. The calculi of the clinical benefit should demonstrably outweigh the contrary. It is important to note that the presence and identification of genetic variations such as a SNP does not automatically imply that the P/RS may be at risk. In fact, some mutations or variations may be totally benign variant or likely benign, likely pathogenic, pathogenic, or variants of uncertain significance within the genome and may probably not have any pathogenic effect and therefore not clinically actionable. For example, changes or mutations in the ATP8B1 gene have been implicated in *benign recurrent intrahepatic cholestasis* (BRIC types 1 and 2).[42] While there are phenotypic expressions of these mutations in the form of liver irritations and jaundice among others, from the clinically perspective, this may not be necessary and actionable per se since the genotype of BRIC is benign based on current scholarships in molecular biology.[43] In this context, confidential information on this may not necessarily be divulged to relatives. However, oncogenes usually have the potential to either overexpress or mutate into tumors and eventually lead to cancer such as BRCA 1 & 2 genes.[44] In addition, suppressor genes typically are responsible for cell cycle regulation and apoptosis but could mutate causing cells to progress to cause certain cancers such as the p53 genes. Indeed, mutations in p53 genes could be inherited and lead to Li–Fraumeni syndrome (LFS) in some people but can also be due to mutations during embryogenesis with a seemingly high probability of causing cancer by age 70.[45] These two genetic polymorphs or variants have the *probability of harm* and the *magnitudes of the harm* have been clinically validated and therefore clinically actionable. Consequently, if a P/RS have any of these genetic profiles, healthcare providers and researchers have the onerous duty to inform them to seek for immediate clinical care. In addition, since these are typically inheritable genetic diseases, they must inform their familial relations to get tested immediately. Should P/RS refuse, healthcare professionals have the duty to disclose such confidential clinical information to potential carriers of these genes for clinical action or intervention. Based on current professional, regulatory, and ethical standards in the United States, early identification and personalized medical intervention is essential in saving lives. Another example worth illustrating here is the alpha-1 antitrypsin deficiency (AATD). AATD is an isolated human growth hormone deficiency (IHGD)—carriers of this genetic variant who smoke have an extremely high propensity of getting lung cancer. In this case, an early identification of potential victims of the genetic variant may not necessarily have any treatment but may instead be counseled about the dangers of smoking because of the *foreseeability of harm*. Disclosure of such confidential information to P/RS' relations are also significant in order for them to make specific lifestyle choices and adjustments in view of the *preventability of harm*. After all, as the axiom goes, prevention is better than cure. Concealing such critical genetic information from carriers of these variants could cause more harm than good.

Another classic example of the breach of confidentiality is the *Safer v. Estate* of Pack case.[46] Safer's father had recurrent polyposis of the colon and this later metastasized into colon cancer and was treated by his physician, Dr. Pack. At that time, the physician purportedly knew that this type of colon cancer was inheritable in the family but failed to warn or disclose to his children their predisposition to the disease. Later, Safer was diagnosed with the same type of inheritable colon cancer (the disease was already at an advanced stage) and sued his father's estate alleging that he should have been warned about the inheritable nature of the disease and would have sought for an early treatment. The New Jersey trial court affirmed Dr. Pack's defense. On appeal however, the Superior Court of New Jersey opined that

> No impediment … to recognizing a physician's duty to warn those known to be at risk of avoidable harm from a genetically transmissible condition…. There is no essential difference between the type of genetic threat at issue here and the menace of infection, contagion or a threat of physical harm.[47]

Accordingly, Safer's suit was affirmed. In other words, physicians have the professional and legal duty to warn and thus disclose confidential genetic and other information to patients as well as their families especially if these are inheritable and clinically actionable. While these legal landmarks seem to be accepted generally, it is worth noting that not all people have accepted the disclosure of confidential genetic information. Some opponents of confidential medical disclosure have argued that informing relatives of risks of possible inheritable or transmittable genetic conditions and disease may cause substantial harm. It has also indeed been documented that some people actually experienced anxiety, alienation, and depression when they are informed of the possibility of them being carriers of some genetic risk factors. Some family members may blame others for imposing diseases on them, which could lead to some tensions. Also, there are many inheritable diseases that are not curable as illustrated above. Consequently, the argument is that disclosure itself causes some modicum of avoidable harm rather than good and after all one of the dictums of biomedical ethics is *primum non nocere* (above all or first of all do no harm)! Disclosures do cause some harm than good. In view of these and a medley of other reasons, some opposed disclosures. Others proposed that a distinction be made between actionable and nonaction genetic disclosures. Since the case for actionability has already been discussed, I wish to examine the issue of nonactionability in the context of disclosures of confidential genetic information.

The presence of SNPs or mutations in genes does not always imply a carrier will automatically manifest as a disease. Obviously and as noted above, there are some genetic variants that call for clinical immediacy but there are some that do not. In addition, certain genetically induced diseases such as Huntington's are not curable but early detection may be helpful in managing the symptoms or other conditions associated with it. Furthermore, iatrogenic diseases, may occur when healthcare providers with preponderant intent to bring about well-being but produces contrary effect.[48] For instance, the use of tamoxifen has been implicated in endometrial cancer.[49] Also, DNA or gene damage can also be caused by exposure to heavy metals such lead, UV rays, ionizing radiations, hydrogen peroxides.[50] Such exposures could damage the p53 gene. Gene damage could also be due to sudden mutations or "medically unexplained symptoms."

Indeed, certain genotypic variant do not always call for any or immediate clinical and therapeutic interventions; in this case, disclosure of such confidential information should not be readily effected. At least proper counseling ought to proceed any intent to disclose weighing the potential of disclosure and no disclosures of genetic confidential information. I believe that the IRB, the attending physicians, genetic counselors, and experts in the particular molecular biological sciences in each medical and research facility be actively involved in the determination of what constitute actionability and nonactionability process. Disclosure of confidential genetic information should diametrically be determined by empirical data to the extent that such information so disclosed could unequivocally be tied in tandem to patients' clinical history with preponderance consequential benefits. I believe this will ensure that in the event of intentional and professional consensual breaching of confidential medical information, there will be convincingly justified reasons that P/RS will be in some tacit agreement with.

As a recapitulation, genomic data belong to the patients, research subjects, and their progenies or families. That is, genetic biodata is a shared social information or entity that many people could justifiably make a claim to and this in turn posit many ethical quagmires. Does a patient have a fiduciary obligation to share or inform relatives of genetic information if he has an SNP with a high propensity for a particular disease such as cancer? Will he be compelled to break that confidential rapport between a physician–patient and inform a "third party"? Does genetic information constitute a public health issue?[51] There have been divergent views on the extent of keeping genetic information confidential. This is because every individual is autonomous. In respecting the autonomy of the individual or patient, it may be argued that such genetic information should not be shared with anyone except the patient being tested. In addition, healthcare providers are required to keep a modicum of confidentiality. Therefore, even if the genetic test contains some potentially debilitating information that may affect other relatives, neither the patient nor his physician has the "right" to share. On the contrary, it may be argued that healthcare providers have some fiduciary obligations and a moral duty to disclose potentially debilitating clinical information even if such genetic information were confidential.[52] In addition, greater confidentiality and protection of genomic data are critical in precision medicine. As the director of the NIH, Dr. Collins rhetorically posited, "Ask people, 'Are you comfortable having this specimen used for future genomic research for a broad range of biomedical applications?' If they say no, no means no."[53] This profound rhetorical question and statement affirms the postulation for respecting the autonomy of the individual to make his/her own decisions about genomic indices as well as ensuring that such data could potentially be useful for society in general if allowed. In brief, disclosure of genetic biodata may only be justified based on the clinical calculi of validity and *actionability* that the benefits accrued outweigh any risks with consent.

Informed consent

The prospects of pharmacogenomics and PM are magnifying seemingly intriguing questions about IC in genomic medicine and research. Due to the fact that many people have identical genetic information, should everyone within a particular group give IC for pharmacogenomics clinical trials and therapeutic procedures? How will such sensitive genetic information be protected or how should the company hold such information in terms of confidentiality? Also, how will a patient or individual be protected when the company is sold or acquired by a third party as in the case with deCode becoming NextCode? This question in particular has ethical and legal ramifications. Should IC in terms of genetic information be limited to or confined to only the company performing the initial test? Recently,

Lars Steinmetz and his astute team of researchers stirred a turbid pool of controversy when they attempted to publish the data of the entire genome of the HeLa cell line for the first time.[54] They did not obviously obtain any consent from the Lacks' family prior to the publications. As anticipated, the attempted publication obfuscated into a morass of ethical issues because the sequenced HeLa genome contained sensitive information that is shared by her surviving progenies; indeed, it is possible to know their genetic traits and predisposition to diseases that are protected by laws such as HIPAA. As a matter of fact, IC was not obtained at the time of her death nor from her family when the specimens were taken from her in 1951 to develop a cell line.[55] As the Executive Director of the Genetics Society of America (GSA), Dr. Fagen noted, "Although NIH played an essential role in the discussions with the Lacks family about the use of HeLa cells, we all need to think about how we approach issues that arise as science moves forward, balancing privacy concerns with advances in research, and the ways policy can be updated to reflect these complexities."[56] There is no doubt that pharmacogenomics companies will need information from different subjects and varying demographical groups for research. In soliciting for such information, an individual may be availing his family's genetic information to the company. The issue of IC becomes even unpredictable because if any family member raises any objection, this can truncate the development of the pharmacogenomics study and this could undermine the prospects for precision medicine. The scope of these limits would also continue to pose ethical challenges and I intend examining these issues in subsequent paragraphs.

First, I believe the question of IC is very important given the challenges indicated above; hence, it is worth examining it in the context of the new frontier of genomic and precision medicine. The term *IC* is at the very core of biomedical ethics and any research involving human subjects. According to the Webster's Dictionary, "consent" (as verb) means to approve, to agree, to comply or yield, or to grant permit—used as a noun, it means approval, agreement, or acquiescence. In addition, "Informed" means to be apprised of or having prepared with knowledge or information. The juxtaposition of these two words implies that research subjects or patients (RS/P) have full disclosure of the extent of a proposed research or therapeutic procedure. These include the objectives, anticipated risks, benefits, kinds of information being sought for the potential use of any data generated among others in the intended research or procedure. A subject giving his or her consent is thus knowledgeable or informed about the research and can at any time withdraw from the research without any impediment or punitive ramifications. IC also involves a high degree of transparency and trust. IC thus recognizes and respects the autonomy of the individual or the group participating in the research. The absoluteness for obtaining IC is significant and legitimizes any credible scientific research and medical procedure involving humans capable of free and IC devoid of any duress or manipulation. In my estimation, IC has a tripartite implication among others—it protects vulnerable subjects, the investigators, and the sponsoring agency or institution. The IC process and documents must be officially approved by an IRB and made available in regulatory depository prior to the commencement of the research or therapeutic procedure in compliance with FDA's *FDAAA 801 Requirements* or *WHO International Clinical Trials Registry Platform (ICTRP).*[57] A failure to obtain IC is a serious violation of both international norms such as the *NC, DOH*, and *Code of Federal Regulations*, Title 45 Volume 46 and US directives among other stipulations. The *NC* in pertinent parts states, "The voluntary consent of the human subject is absolutely essential."[58] In addition, the DOH also recommends that research subjects must be properly informed of the aims, sources of funding, potential conflicts of interests, institutional affiliation (school, hospital, government agency, etc.) of the researcher. Furthermore, if the subject understood the essence of the research as encapsulated in the consent form, then the participants should concur and affirm either written or non-written and witnessed under the expediency of the exculpatory clause to be able to withdraw at any phase of the research devoid of any ramifications.

Furthermore, FDA rules 21 CFR 50 imposes a requirement on researchers to obtain IC and rules 21 CRF 56 in pertinent part requires an IRB review. In rule 21 CFR 312.66, researchers are entreated to assure subjects of their assiduous preparedness to obtain an IRB review that includes indications that IC has been obtained prior to commencement of a research or therapeutic procedure. These three rules are in essence similar to the Common Rules 45 CFR 46. However, unlike the Common Rules, they focus on ensuring safety and efficacy of FDA approved products for the general public rather than on biomedical research. Nonetheless, each of these requires IC in anticipation of a greater protection of vulnerable human subjects. HIPAA governs protected health information (PHI) of each patient. Patients may so authorize the release of their PHI for research purposes. That is, give consent; typically a written consent to the researcher for specific purposes only. The researcher cannot use it for any other purpose other than what has been defined in the release document or consent form. Obviously, HIPAA regulates the health insurance industry in a bid to protecting patients' health information, mitigating fraud, and governs the process for consenting for the disclosure of information by authorization while the Common Rules (45 CFR 46) and the FDA Rules (21 CFR 50) ensure that RS/P give proper consent to researchers. These rules are generally applicable to genomic data because they also constitute clinical data capable of identifying individuals. Consequently, researchers are obliged to obtain IC from their patients for the use of their biospecimen or genetic data. The litmus test of this was in the *Moor v. UCLA* case. His attending physician used his biospecimen to develop a cell line without a written consent. In fact, he *consented* and gave his specimens including tissues and blood to Golde but was not explicitly and adequately informed that these were going to be used for other research and commercial purposes. While a detailed analysis of the case has been discussed in the second section of this book, suffice it to say that this case illustrates the potential consequences of tacit violations of the rules governing IC. The court indeed found that there was a breach of Moors rights for a failure to obtain IC from him. Another case worth discussing is *Greenberg et al. v. Miami Children's Hospital*. Greenberg group and a number of not-for-profit organizations entered into a collaborative agreement with Dr. Matalon to identify and study the specific gene loci responsible for Canavan disease (a debilitative degenerative disease with no known cure).[59] Because of the support of these organizations, many laboratories offered the test pro bono. However, Matalon and his team patented the process for the gene test without an IC from the organizations and also, laboratories and hospital could not offer the test pursuant to the patent act without permission. While the court found that Matalon and collaborators indeed benefited from royalties paid to them in view of the patent, they did not find that there was a breach of IC within the context of the PPRs. As some scholars have noted, IC should not have exculpatory language that may be advantageous to the researcher.

Furthermore, research involving genetic materials or biospecimen is unique and presents labyrinth of ubiquitous ethical quadrants. This is because close genotypic affinities of familial genetic data, the hitherto IC generally accepted have increasingly been subjected to analysis and debates. Simply put, an individual's consent to participate and donate his biospecimen for research poses the existential risks of exposing his relative's genetic information such as genomic variants, disease proclivities among others without their expressed written consent into the hands of researchers who might not necessarily know the family but nonetheless may have access to substantial information that could harm them. Practically, it is impossible to seek consent (not even proxy consent) from *all* relatives in order to conduct research. In addition, IC in *GWASs* can be tangentially complicated.[60] This is because such GWAS could involve many research subjects transcending many socio-demographic and geographical locations. GWAS are significant in the development of pharmacogenomics and PM. For example, through GWAS, researchers have mapped the genetic factors implicated in diseases such as types 1 & 2 diabetes and Crohn's disease.[61] Genomic data from such studies may have clinical values to people that might not have participated in the research. Researchers may be faced with the hurdle of publishing such information. But is it feasible and possible to obtain IC from all people who might share the same gene of interest or biomarker? While the obvious answer theoretically, may be in the affirmative, pragmatically, these seem impossible. However, concealing genomic data can and does lead to duplicity and redundancy in research in addition to increasing costs and potentially putting some population at risk. IC thus has some limitations in view of these challenges. The NIH has established a data collection portfolio (NIH Genomic Data User Code of Conduct) on GWAS and has defined clear directives on how to access such information, who accesses these, and sets limit on how these should or should not be published. While these directives seem convincingly clear, the technology and the security associated with these and the extent of vulnerability such as potential for clandestine access will ultimately be determined by how safe and sustainable these will be adhered to.

IC is even critical in genomic medicine due to the fact that genetic data accrued from the research may be useful and applicable to other members of society that might not have even participated in the research. The complexities of consent during genomic research cannot be underestimated. Recently, the Havasupai Tribe participated in a genomic study on the causes of type II diabetes in which they gave their biospecimen to researchers from Arizona State University.[62] Members of the tribe purportedly have a high incidence of type II diabetes, and so researchers wanted to study the possible genetic component to the disease. After many years of laborious efforts, researchers could not determine the genetic underpinnings to the disease contrary to their expectations. However, researchers then decided to use the biospecimens collected for *other studies* without due diligence in obtaining their IC after the original research was truncated. As anticipated, these degenerated into medicolegal, socio-anthropological, and ethical discourses. Did the researchers actually and intentionally refuse to obtain IC? As it turned out, researchers were surprised because they did not see the need for a new consent even though they settled the case out of court.

Another ethical issue worth discussing is the risk of information leaks especially posthumously with malicious intents. This is an increasingly emerging area in biomedical research. In fact, Declaration of Geneva prohibits the disclosure of medical information upon the death of a patient. That is, the right to be forgotten posthumously and confer some sense of dignity. Obviously, obtaining IC may not be feasible unless there is a designated person to do so. In terms of genetic information, it could be challenging because relatives may still be alive who might either want or not consent to the use of genomic data. Also, genetic data or material might need authentication if there is the modicum of suspicion of cell-contamination or there is the need to undertake comparative study.[63] Often cells or tissues get contaminated; for instance, with fibroblasts or mixed with other biospecimen of unknown sources to the extent that researchers often felt the need to authenticate their cell lines. For instance, HeLa cells have been used so much and have been cultured in many laboratories with various media and techniques over the decades and increasingly, researchers are finding out that their samples have been contaminated. One of the most effective ways was to contact some members of the Hela family in order to have their biospecimen analyzed and used as control for confirmation—sometimes some protein sequence of interest may be amplified from the original cell lines for comparative analysis or GWAS. As intimated above, it calls for re-consenting from the familial relations of original donors in posthumous situations or if the original donor might not be readily located. As the fifth article of the *Declaration of Geneva* states, "I will respect the secrets that are confided in me, even after the patient has died."[64] This could be a potential hurdle for researchers in genetics especially if a patient explicitly does not allow any future use of his medical information. But Federal Regulations 45 CFR 46.116 (b) (5) and 21 CFR 50.25 (b)(5) allow the possibility for re-consenting if "significant new findings developed during the course of the research which may relate to the participant's willingness to continue participation will be provided to the participant." In terms of genetics, I believe relatives with shared genetic information may *re-consent* (even if the original consenter passes away), especially if the research is going to be clinically applicable to them or confer some therapeutic benefits to society in general. As in the case above, the Hela family eventually *consented* for their specimen to be used decades after she passed away. Some schools of thoughts are of the view that researchers should include in their original consent form, specific allowance for participants to name someone to re-consent on their behalf posthumously or alternatively categorically preclude the future use of their biospecimen.[65] In brief, consent and re-consenting are important in genomic study. Since most genetic information is clinical data, it is significant to obtain full consent from participants or patients. In addition, because genotypic data are a shared data, the issue of social sensitivity should be considered all the time to protect vulnerable people or patients who might be directly or indirectly participating in the research. Since it is impossible to obtain IC from every member of GWAS, utmost privacy must be guaranteed for those who consent to participate.

In addition, there have been increasing mergers of many biotechnology companies over the past couple of years. As a matter of fact, over 1300 pharmaceutical companies merged or were acquired between 1999 and 2009.[66] While this phenomenon is not new per se, nonetheless, it has implications on the development of pharmacogenomics in particular and irks some ethical issues as well. While details of these mergers and acquisition might not be known since most are shrouded in commercial contract laws, it has some implications for bioethical and regulatory considerations. For instance, what happens to genomic biodata? While they are protected by privacy laws *in situ* especially in the United States and United Kingdom, there is no guarantee that sensitive private data are in reality protected. Also, this can be challenging if these acquisitions involve multinational pharmaceutical companies. One of the examples, *locus classicus,* involves a genome sequencing company called deCode, which was founded originally in Iceland by Kari Stefansson in 1996.[67] The company was acquired by Amgen in 2012.[68] Later, NextCode emerged out of deCode as a new company and offered virtually the same genomic testing products and services. Just at the threshold of this book, the baton of proprietary ownership of the company has been acquired by Wuxi Pharmatec (Wuxi NextCode) in 2015. Ethicists and the regulatory agencies are watching as to how genomic data or gene banks will be used by the new owners.

In view of these challenges, some schools of thought have hypothesized that researchers should be discouraged in the use of exculpatory language, which might shield them from any ramifications if there is new ownership of the company. Indeed, some scholars are of the view that provisions be made to allow for new consent (re-consenting) each time a company's operational portfolio is transferred to a different person or sold to a third party. Re-consenting will obviously ensure greater protection of data and prevent potential misuse or disuse of private genomic data or information. It is further postulated that regulations allowing for the release of personal medical data under HIPAA should take prominence, allowing research participants to exercise their *privacy decisional capacity* guaranteed by HIPAA to ensure and insure that their genomic data are protected at all times. If the company is sold, the new entity must request for a new consent approved by an IRB or as so defined in current local and international regulations before the deal for the merger or acquisition is completed. In addition, in the consent process, participants should have the ability to limit the use of data to specific geographical locations. The DOH explicitly states that owners of genetic materials and data have the rights to opt out or request that their genetic samples be withdrawn at any time.[69] Since gene banks or genetic data are not the preserve of only an individual, it calls for international coordination and involvement in monitoring how they are used by third parties or posthumously. IC, however, is not necessarily an international concept per se. In Chinese medical tradition, some scholars have noted that "consent" may be obtained from relatives of competent adults. As Xiaomei Zhai once noted in a seminal paper addressed to the Asian Bioethics Conference,

> In traditional Chinese culture, greater moral meaning and values rests in the interdependence of family, which transcend self-determination. One of the distinguished characteristic of Chinese culture is that they are often less individualistic than those in Western Europe and South America. Therefore obtaining informed consent from the spouse or family member instead of merely from patients themselves is a conventional procedure in the medical practice in China.[70]

One wonders how Wuxi NextCode will handle issues of voluntary IC in the new company. If they follow current practices in China, for example, it could degenerate into some amorphous situations for clients with ethical implications. It is in anticipation of challenges such as these that some scholars call for time limits on how long biobanks could store biodata of their clients without a new consent. This book postulates a distinctive consenting paradigm that confers greater protection of genomic data and at the same time guarantee the ever-expansive watershed of PM during the merger of transnational biopharmaceutical companies. By this, I mean a kind of transitional caretaker laws/guidelines. In this perspective, biodata and stored biospecimen such as tissues will be transiently protected by existing medicolegal and ethical norms of the genomic donors' country of origin. For example, in the United States, it will be HIPAA and the Federal Guidelines that automatically take precedence during the transition period of the company to new owners. Upon successful transition, the new owners/management will then re-consent with applicable norms in order to access or use biodata of client's information acquired during the merger. This will confer a greater protection as well as respect existing local norms.

In perspective, genetic information and biospecimen must not be treated as an ancillary medical data. Indeed, the cost of sequencing whole genomes has dropped abysmally from over $1 billion to as low as $1000 with a high propinquity that it could even be lower; it will be expedient to integrate genomic testing as normal clinical routine analogous to blood/fluid tests typically required upon a doctor's visit. Written consent must be sought from each patient or research participant. In addition, data extrapolated from genomic testing are readily protected by existing regulations.

Absolute protection of genomic data will encourage people to participate in future research and needless to say, this is critical in the development of PM. Indeed the Ethical, Legal, and Social Implications (ELSI) department within the NIH was established to study these emerging challenges. Finally, as the Presidential Commission on the Study of Bioethical Issues noted,

> Not unique to whole genome sequencing, a well-developed, understandable, informed consent process is essential to ethical clinical care and research. To educate patients and participants thoroughly about the potential risks associated with whole genome sequencing, the consent process must include information about what whole genome sequencing is; how data will be analyzed, stored, and shared; the types of results the patient and participant can expect to receive, if relevant; and the likelihood that the implications of some of these results might currently be unknown, but could be discovered in PRIVACY and PROGRESS in Whole Genome Sequencing and the future. Respect for persons requires obtaining fully informed consent at the outset of diagnostic testing or research.[71]

In brief, IC is a core mandate of genomic study especially as applicable to precision medicine. Any violation of this could lead to several consequences such as potentially discouraging participants and increasing the potential risk of infringing on the rights of vulnerable subjects to mention just a few. IC should be dully obtained at all times and at every level involving human subjects in view of developing a robust pharmacogenomics and PM.

Genetic stigmatization and essentialism

There seem to be a truism in the assertion that the genome is a paradigm of the human person. In other words, "There is no normal genome that is expressed in a 'normal' person."[72] Furthermore, a human person is not just made up of genetic codes; the environment plays a crucial role as well. Paradoxically, as indicated above, there is no person devoid of any genetic variant; it seems that the norm of human genetic information is that each person has some genetic variant or uniqueness either spontaneously or as a result of many factors. Each individual is shaped as a result of the intricate interplay of culture, environmental factors, geophysical location, and lifestyle choices and other socio-dynamic factors. But there seem to be a reductionist tendency to rely solely on genes as the underlining cause of diseases and human behavior with little or no emphasis on the roles of the environments and social interactions. The consequences of these are manifold. Excessive reliance on SNPs may perniciously lead to genetic essentialism analogous to the adverse impacts of the eugenics movement. It could further degenerate into situations where patients could be compartmentalized according to their genetic architecture to the exclusion of other essential factors in their lives. Genetic essentialism could lead to racial fragmentations and discriminations as some demographic groups may share certain "adverse" SNPs that predispose them to certain diseases and potentially to unfathomable ridicules.

Genetic stigmatization remains one of the controversies and perhaps unavoidable snags in the development of pharmacogenomics and PM because genes are biological identifiers. Despite the fact that humans as biologic entities have enormous common or similar identifiers such as our anatomical structures, biochemical components or constituents, neurological and pharmaco-nutrients synthesis and pathways, there are obvious differences or identifiers that manifest in physical differences and at the genetic levels typified the presence of SNPs and others. Some of these differences and similarities have been shaped by the process of evolution and the process of external factors. Indeed as Dobzhansky insightfully noted in his famous essay, *Nothing in biology makes sense except in the light of evolution.*[73] Evolution often leaves indelible marks on individuals and collectively as species evidenced in our differences and stark similarities as humans. Some people are born short or tall and others have the genetic and environmental proclivities toward obesity or other diseases they might not have control over. There are phenotypic manifestations of certain good SNPs such as in sickle cell anemia (SCA) where carriers of this genetic variant have just one mutation—glutamate is replaced by valine in the hemoglobin chain. Because of this single mutation, carriers may have relatively shorter lifespan compared to the general population but have better tolerance to malaria.[74] This is critical for populations in the tropics that tend to have most of the clinical incidence of malaria in the world. Early identification of such a gene could help clinicians offer better personalized care and improve the quality of their lives even if they have a shorter lifespan. But it could also lead to stigmatization within the population. Another example worth mentioning is hemophilia, a rare genetic variant manifested in the failure of blood to clot. This could be clinically fatal as patients could bleed to death because their blood may not clot normally. Another genetic disease worth mentioning here is *testicular feminization syndrome* (TFS) or *androgen insensitivity syndrome,* which is reputed to occur in 1 out of every 65,000 male births.[75] While carriers of these genes are chromosomally males, phenotypically are females with female genitalia. The manifestations of these genetic variants could lead to stigmatization in certain cultures or societies or subgroups. Such stigmas could be incorporated into the very fabric of the larger society and could isolate carriers of these genes and even complicate their clinical and physical needs instead of care and proper management. Given the above examples, it is worth expatiating further on the concept of "stigma" and its implications for PM.

Etiologically, stigma has its roots in the Greco-Roman culture where the word was generally used descriptively for tattoos (typically on the skin of servants or slaves), or marks on other people as a sign for the general population to be weary of them or stay away from them because of perceived dangers among others. The concept has metamorphosed and developed in our cultures even today. This etiological basis of the word seems to form our contemporary understanding of the word stigma. In a recent study on this concept, Corrigan in a recent paper used this framework of the word to enumerate the components of stigma. She suggested that "Stigma marks someone as a potential target of negative reactions," where "the stigma prompts others to apply negative stereotypes, cognitive frameworks that give meaning to signals," and these "stereotypes contribute to affective response such as fear, pity," and these "affective responses may escalate into discrimination against members of the stigmatized groups such as social avoidance."[76] In addition, some scholars have also pointed out that …because group living is highly adequate for human survival and gene transmissions, people will stigmatize those individuals whose characteristics and actions are seen as threatening or hindering the effective functioning of their group.[77] These subtle assertions have symptomatic imports for genetic stigmatizations and health care because a genetic uniqueness could constitute a stigma for an individual. Furthermore, stigma is a potent stressor that can manifest itself in psycho-neurological forms and by itself does affect the health of individuals labeled as such. Studies have consistently and convincingly shown that stigmatization of people with certain unique identifiers such as genetic, diseases, and behavioral patterns invariably have effects on their health and the quality of their lives in general. In certain cultures or societies, people with certain physical identifiers (that are often caused by genetic variants) such as *androgen insensitivity syndrome* albinism often suffer social isolation and humiliations. As Jenerrette insightfully noted, "*Stigmatization* is the process of identifying an attribute of a person or group and associating the attribute with a stereotype that negatively labels or brands another in a way that is perceived as disgraceful by society. More specifically, health-related stigma refers to a form of devaluation, judgment, or social disqualification of individuals based on a health-related condition."[78] In a seminal and pioneering work, Goffman also indicated that a stigma is an "attribute that is deeply discrediting" reducing the person who possesses it as "from a whole and usual person to a tainted, discounted one."[79] Sources of such "stigma" include individuals with "abomination of the body, blemishes of individual character" and tribal character (in my estimation, racial or ethnic stigma). As Scambler also noted, stigma occurs when "people to whom a stigma is attributed are imperfect beings possessed of putative defects that is beyond their capacity to correct."[80] Stigmatization is accusatory because patients with some form of diseases are negatively stereotyped typically with pejorative accolades that could potentially dehumanize them. Mounting evidence seems to be emanating from several studies about the impacts of genetic stigma on the overall quality of life and health of individuals or subgroups with genetic identifiers. Such studies have also some leaning to the assertions that stigmatization could truncate the development of PM since potential human subjects may decline to enroll in crucial genome-wide studies (GWS) that may otherwise have impact on our understanding of the genetic underpinnings and pathways of diseases. Indeed, according to Dar-Nimrod and Heine, "Research indicates that perceiving genetic causes of a characteristic or behavior is associated with deterministic thinking."[81]

For instance, Sophie Lewis et al. also conducted some studies on the impact of stigmatization of obesity on health and noted with such chagrin:

> Exposure to stigmatising attitudes and behaviours also prevented participants from taking part in activities that would improve their physical health and wellbeing. Participants described how the combination of direct, indirect and environmental stigma prevented them engaging in exercise in public spaces. Some stated they were unwilling to participate in [the] exercise because they "expected" that people would "laugh at," "ridicule" "stare at" or "abuse" them. One participant (a 34 year old female) said that she rarely participated in physical activity, because she felt constantly "on display."[82]

Stigmatization of people with obesity, for example, often undermines the significant roles of genes as causal agents for the phenomenon. And even seriously, such social attitudes do impede the clinical intervention and management of obesity and needless to say it could pose policy challenges as well because stigmatization sometimes blemishes the individual into a tainted person.[83] As indicated above, genomic studies help identify specific genes of target for therapeutic purposes. Some of the benign and unintended implication is that genomic study for SNPs or biomarkers could easily huddle some people with certain genetic variants into categories in terms of their susceptibility for certain diseases such as cystic fibrosis, HD, SCA, and Canaan disease. These genetic taxonomies or "gene pools" are helpful in tailoring therapeutic interventions. On the other hand, some studies have also shown that genetic stigma or genetic framing of diseases could help some patients to recover or opt for treatment if available. As Michele Easter pointed out, "There is evidence that genetic framing is helpful for countering stigma in eating disorders, despite findings to the contrary for other mental illnesses...."[84] But historic antecedents have demonstrably proven that genetic screenings were used to identify and *ostracize* people with some genetic variants from *healthy members of society*. As Pamela Sanka et al. have also noted, "The historical link between genetics and eugenics might account for heightened fears, and the frequent reliance in related commentaries on examples of serious, even fatal, conditions such as Tay–Sachs, might subtly contribute as well to the belief that genetic conditions are inherently stigmatizing."[85] In addition, genetic screening became a razor edge for social policies such as the eugenic movements in the twentieth century in Europe and the Americas where lives of many innocent people were violated with impunity.[86] For example, the *Eugenics Record Office* (ERO) officially used seminal works in the study of genetics to fuel and orchestrated formidable public policies such as forced sterilization and restrictions on marriage of epileptics with the preponderance intent to purge the population of people they deemed (though erroneously) with undesirable traits.[87] In a word, the *ERO* restricted the propagation of the genetically "unfit" and so formulated and influenced laws that were aimed at eviscerating the *bad germplasm* in society. By 1931, it was estimated that over 13 states had promulgated such sterilization laws resulting in nearly 15,000 people being sterilized because of their genetic dispositions and manifestations of diseases that were considered perilous to society at the time.[88] These sentiments were stridently expressed by Henry Fairfield Osborn during the *Second International Eugenics Movement* in these words:

> In the United States we are slowly waking to the consciousness that education and environment do not fundamentally alter racial values. We are engaged in a serious struggle to maintain our historic republican institutions through barring the entrance of those who are unfit to share the duties and responsibilities of our well-founded government. The true spirit of American democracy that all men are born with equal rights and duties has been confused with the political sophistry that all men are born with equal character and ability to govern themselves and others, and with the educational sophistry that education and environment will offset the handicap of heredity.[89]

While we might look with disdain at such unfounded and demagogic claims about genetics today, nonetheless, it influenced public and medical policies for decades. We see categorical racial discriminatory overtones in these statements and the purported "scientific papers" presented at these conferences. Genetic *projectionism* or the reduction of a person to the essence of his genetic or hereditary information was a violation of every known human ethos and had obfuscated and eve truncated clinical judgments at the time. It is important to note here that the Eugenics movement was a global phenomenon to the extent that some countries such as Denmark, Germany, Finland, England, and other countries also passed laws that were used to exterminate people with genetic variants. The most systemic and brutal of this was the Nazis where unfathomable clandestine medical experimentations resulted in the loss of thousands of innocent lives. Such brutalities under the aegis of genetics and medical research have left an indelible mark on the science of genetics, eugenics, and society and it is impossible to be ignored even in contemporary study of genomics and precision medicine. Overt emphasis on the role of genetics gives credence to mechanistic insights of the polygenetic variants of genetic architecture of diseases and egregiously undermined the complexities of the human person and the role of nurture and other socio-environmental indices in human development.[90] Today, more than ever before, genomic research and the emergence of precision medicine are under the aegis of strong legislative and regulatory aperture. Many laws, policies, and specific ethical guidelines continue to shape genomic scholarships both locally and internationally. While the enthusiasm and the thrust for precision medicine seem formidable and popular, it is important to tread cautiously especially in the protection of vulnerable populations. Population-based genetic study should adhere to the highest medical ethos of confidentiality, IC, autonomy, and privacy especially in an increasingly information-driven world. As one scholar cautionary noted:

> The Human Genome Project and newly developed genetic information offer us a potential for fabulous medical and social advances—a

> chance overcome problems that have cursed our species from time immemorial. They also offer us the opportunity to perpetuate the worst that our species has developed over centuries, and to develop a true caste system. We ought to design the legal and social institutions that will control the use of genetic information with the presumption that intellectual excitement associated with genome research . . . will provide powerful incentives for both paths.[91]

Furthermore, access to health care is linked partly with employment and health insurance companies, especially in the United States may try to mitigate costs and therefore could use pre-enrollment information and screenings in order to calibrate premiums. Genetic screening (if required prior to enrollment or at any time after enrollment) could potentially unearth certain biomarkers that may place some people at a higher risk of certain diseases such as cancer. This could lead to a higher premium or denial of health insurance coverage (even though prohibited by law). Also, employers could use genetic screening to glean information of potential employees in order to offer certain job-related accommodations and disability benefits to them. While this might appear novel (albeit benign), it nonetheless could lead to discriminatory practices to the extent that health care could *essentially* be based on genetic information to the exclusion of crucial socio-environmental variables such as diet, education, and personal hygiene just to mention a few. The concern is that genomic information could potentially be used by health insurance companies and employers to either overtly or covertly discriminate against employees and enrollees. This issue was acknowledged during the *HGP* and so the *Genetic Information Nondiscrimination Act* (GINA) was formulated as a guide for researchers and stakeholders. Recently, Fabricut was found guilty (Civil Case No.: 13-CV-248-CVE-PJC) of violating GINA when it refused to hire one of their temporary employees for a permanent position because during medical screening, she was purportedly diagnosed with carpal tunnel syndrome (CTS) a neurological disorder caused partly by genetics and other etiological factors that remains idiopathic. This is troubling because this lawsuit occurred (May 7, 2013) after GINA was promulgated. Unlike other guidelines, such as the DOH, *NC*, the *Belmont Report*, genetic guidelines are fragmentary as there are many of them at both the federal and state levels and internationally. There is the urgent need to specifically codify these into single comprehensive documents that unequivocally addresses some of the challenges in the area of genetic discrimination. It is worth mentioning that the *NIH Genomic User Code of Conduct* gives clear and definitive standards and requirements for accessing and using *Genotypes and Phenotypes* (dbGaP) database in pharmacogenomic research.[92] My contention is that this document could be expatiated and become *sine qua non* policy for international genome-wide study since it ensures even greater protection of individual genomic information.

Genetic testing and the promise of therapeutic interventions is creating lots of prospects within the scientific community. This and many other factors seem to be excellent launching pads for researchers to collect and analyze genetic materials across the globe and especially from native tribes or groups. Some of these groups include the indigenous tribes in the Amazon regions of Brazil, the Amish, and many Native American generally considered vulnerable population in the genre of bioethics. While the search for genetic material and data extrapolated from these hold great prospects for PM, there are some perturbing issues emanating from these such as genetic tourism.

It is truism that a tsunami of researchers trouping to vulnerable populations and some think it is just for the curiosity of it akin to tourism. There is generally a renewed interest among researchers to genotype homogenous populations such as the Amish people and some indigenous or Native Americans. This is significant because such genomic data help in comparing other genetic data from other places in order to identify and study population differentiated biomarkers and some genes of interests that might have undergone some evolutionary changes, mutations, or insertions in other heterogeneous populations. These can be clinically significant in developing specific pharmacogenomics interventions. In addition, such homogenous study can and does lead to the identification of some genetic diseases unique to these homogenous groups. For example, some researchers have discovered that mutations in VPS13B gene causes malfunction of the proteins from it leading to Cohen syndrome. But among some Native American tribes, such genetic studies have generated unprecedented controversies. In a seminal work, Tall Bear discusses the socio-historic contexts of the controversies.[93] She observed that DNA testing irks and stirs intra-tribal debacles of the purity, identity, and the old scare of colonization and the continual debates of ownership of their "land" and "nations." Furthermore, some natives feel DNA testing especially postmortem are invasive and disrespectful. As Nick Tipon poignantly notes:

> These are questions that anyone who gives their genetic material to scientists has to think about. And for Native Americans, who have witnessed their artifacts, remains, and land taken away, shared, and discussed among academics for centuries, concerns about genetic appropriation carry ominous reminders about the past. I might trust this guy, but 100 years from now who is going to get the information? What are people going to do with that information? How can they twist it? Because that's one thing that seems to happen a lot.[94]

Therefore, genetic testing might seem to natives as some kind of tourism in which researchers are just visiting them to explore their genetic materials and not necessarily for any other reason that might be of value to them. It should also be noted that visits to indigenous people carry an additional risk of transmitting diseases that might be new to them, since they might lack immunity having never been exposed to germs on the visitors or researchers. For example, "contact" with native tribes such as the Nahua tribe is believed to have led to their decimation.[95] My contention is that any contact with natives especially those that might not have had any or significant contact with the outside world for their genetic materials should be sparingly done. As mentioned above, there are renewed interests in homogenous populations for their genomic information due to their potentials for unique biomarkers that might be valuable for pharmacogenomics development and precision medicine; these ought to be done with utmost care rooted in an impeccable cultural sensitivity.

Also, genetic testing and pharmacogenomics research may be brewing false hopes and as a consequence vulnerable people may be undertaking tests and risky gene therapies while some might even travel to countries with seemingly amorphous regulatory oversights for gene therapy with attendant fatalities. For example, Jesse Gelsinger suffered from *Ornithine transcarbamylase* (OTC) deficiency a genetic disorder caused by the mutations of the OTC gene. He participated in clinical gene therapy study but tragically passed away due to complications from the gene therapy.[96] This and some other challenges have led to stricter regulatory oversights of gene therapy both in the United States and Europe. Currently, the United Kingdom is the only country that has allowed any form of gene therapy.[97] Because of such tough regulatory bottle necks in the United States, many patients interested in the therapy have been seeking for genetic therapies abroad, places with generally weaker regulatory framework in ensuring the safety and efficacy of such therapies. It is, however, anticipated that such therapies may soon be in the United States. Dorothy Romanus et al. assert that "ensuring patient access to said breakthrough therapies through lower cost sharing is key. As evidence evolves and testing for a wider range of known mutations…enters routine care, it will be increasingly important for future economic analyses to consider multiplexed testing for multiple mutations in tandem to fully appreciate the value of personalized treatment in this disease."[98]

As a way of conclusion, the new paradigm in pharmacogenomics and PM has been fraught with many ethical issues with legal and policy implications as well. As a prominent scholar Carl Schneider once noted, "Law provides a rich language for thinking about bioethical issues and is a tool for action as well as talk. But the language of the law, often inapt, regularly fails to achieve its desired effect…. Inevitably, the spirit of the law has penetrated into the bosom of bioethics."[99] To this extent, I will discuss some of the ethical issues such as the question of patentability and ownership of genomic biodata through the nexus of policy and the law.

End notes

1. David Nelson et al. *Lehninger: Principles of Biochemistry* (W. H. Freeman and Company; New York, 2008): pp 322.
2. www.ncbi.nih.nlm.nih.gov.
3. http://ghr.nlm.nih.gov/handbook/genomicresearch/snp. It is important to note that most SNPs do not have any a concomitant effect on the health of the individual per se. But it becomes significant when implicated as the basis for a disease.

4. Ibid. See also, Desmond Nicholl. *An Introduction to Genetic Engineering* (Cambridge University Press; London, 2008): pp 197–199.
5. www.SNPedia.com.
6. Kalus Lindpaaintner. Pharmacogenomics and the future of medical practice, *Journal of Molecular Medicine* 90(1): 2003, 141–153.
7. Matthew Decamp et al. Pharmacogenomics: Ethical and regulatory issues, ed. Bonnie Steinbock, *The Oxford Handbook of Bioethics* (Oxford University Press; London, 2009): pp 536.
8. ⁱNational Research Council: Committee on a Framework for Developing a New Taxonomy of Disease. *Toward Precision Medicine: Building a Knowledge Network for Biomedical Research and a New Taxonomy of Disease* (The National Academies Press; Washington, DC, 2011).
9. Scott Robert et al. *Greek-English Lexicon* (Oxford University Press; London, 2007).
10. Plato. *Politeia* 369b and Aristotle *Politics* 1280b.
11. Ibid.
12. Kant calls this heteronomy as opposed to autonomy.
13. Immanuel Kant. *Groundwork of the Metaphysics of Morals* (1785).
14. John Stuart Mill, *On Liberty.*
15. The Belmont Report.
16. Onora O'Neil. *Autonomy and Trust in Bioethics* (Cambridge University Press; Cambridge, 2002); C. McKenzie et al. *Relational Autonomy: Feminists Perspectives on Autonomy, Agency, and Social Life* (Oxford University Press; New York, 2000).
17. T.L. Beauchamp and J.F. Childress. *Principles of Biomedical Ethics* (Oxford University Press; New York, 2001). 5th edition.
18. Detail analysis of this case will appear in the legal section of this book!
19. Henk J. ten Have. Genetics and Culture: The geneticization thesis, *Medicine, Health Care and Philosophy* 4: 2001, 295–3014.
20. Ibid.
21. Evangelos Evangelou et al. Meta-analysis methods for genome-wide association studies and beyond, *Nature Reviews Genetics* 14: June 2013, 379; also McCarthy, M. I. et al. Genome-wide association studies for complex traits: consensus, uncertainty and challenges, *Nature Reviews Genetics* 9: 2008, 356–369.
22. Danilo Grunig Humberto Silva et al. Genetic and biochemical markers of hydroxyurea therapeutic response in sickle cell anemia, *BMC Medical Genetics* 14: 2013, 108; Jacklyn Quinlan et al. Genomic architecture of sickle cell disease in West African children, *Frontiers in Genetics* 5: February 2014.
23. Chip Walter. A little privacy, please, *Scientific American* July: 2007, 92–95; see also Laura DeFrancesco. To share is human, *Nature Biotechnology* 33: August 7, 2015, 796–800.
24. DeFrancesco. *To Share is Human.*
25. B.A. Binzak. How pharmacogenomics will impact the federal regulation of clinical trials and the new drug approval process, *Food and Drug Law Journal* 58: 2003, 103–127.
26. www.med-ed.virginia.edu/courses/rad/confidentiality/1/oath.html.
27. Ibid.

28. Anneke Lucassen et al. Confidentiality and serious harm in genetics—Preserving the confidentiality of one patient and preventing harm to relatives, *European Journal of Human Genetics* 12: November 12, 2003, 93–97; Jeantine E. Lunshof et al. From genetic privacy to open consent, *Nature Reviews Genetics* 9: May 2008, 406–411.
29. Michael H Kottow. Medical confidentiality: An intransigent and absolute obligation, *Journal of Medical Ethics* 12: 1986, 117.
30. WMA Declaration of Geneva—World Medical Association.
31. Chapter 4: Living to 120 and Beyond: Americans' Views on Aging, Medical Advances and Radical Life Extension (August 2013).
32. Public Attitudes Toward Medical Privacy (2000), available *www.forhealthfreedom.org/Gallupsurvey*
33. www.gpo.gov/USCODE.
34. Kenneth Offit et al. The "duty to warn" a patient's family members about hereditary disease risks, *JAMA* 292(12): 2004, 1469–1473. See also, Elisabeth Ihler, MA. Genetic information and competing interests, *JAMA* 290(9): 2003, 1216–1216.
35. AMA. *Opinion 2.131—Disclosure of Familial Risk in Genetic Testing.* See also, Kenneth Offit et al. The "duty to warn" a patient's family members about hereditary disease risks, *JAMA* 292(12): 2004, 1469–1473.
36. Ibid. Kenneth Offit et al. The "duty to warn" a patient's family members about hereditary disease risks.
37. Ibid.
38. Kenneth Offit et al. The "duty to warn" a patient's family members about hereditary disease risks.
39. *The American Society of Human Genetics Social Issues Subcommittee* on Familial Disclosure Professional Disclosure of Familial Genetic Information 1998.
40. Ibid. This will be expatiated in subsequent paragraphs in detail.
41. *Ethical issues in Patient Safety Research—World Health ....* http://apps.who.int/iris/bitstream/10665/85371/1/9789241505475_eng.pdf
42. http://ghr.nlm.nih.gov/condition/benign-recurrent-intrahepatic-cholestasis. See also, Michael Trauner et al. Molecular pathogensis of cholestasis, *The New England Journal of Medicine* 339: October 1998, 1217–2027.
43. Ibid.
44. Ibid.
45. Sining Chen et al. Prediction of germline mutations and cancer risk in the Lynch syndrome, *JAMA* 296(12): 2006, 1479–1487.
46. Kristin E. Schleiter. A physician's duty to warn third parties of hereditary risk, *AMA Journal of Ethics* 11(9): September 2009, 697–700.
47. Ibid.
48. Iatrogenic risks of endometrial carcinoma after treatment for breast cancer in a large French case-control study. Fédération Nationale des Centres de Lutte Contre le Cancer (FNCLCC).
49. Deborah Tolmach Sugerman. Tamoxifen update, *JAMA* 310(8): 2013, 866.
50. D. Branze et al. Regulation of DNA repair throughout the cell cycle, *Nature Reviews Molecular Cell Biology* 9: 2008, 297–308; Lodish, H. et al. *Molecular Biology of the Cell* (Freeman; New York, 2004). 5th edition.

51. Mepham p 133 See Klitzman, R. How IRBs view and make decisions about consent forms, *Journal of Empirical Research on Human Research Ethics* 8(1): 2013, 8–19 and Disclosures of Huntington's disease risk within families: Patterns of decision-making and implications, *American Journal of Medical Genetics Part A* 143A(16): 2007, 1835–1849.
52. L.S. Lehmann, J.C. Weeks, N. Klar et al. Disclosure of familial genetic information: Perception of the duty to inform. *American Journal of Medicine* 109: 2000, 705–711.
53. Ewen Callaway, Deal done over HeLa cell line, *Nature* 500: August 8, 2013, 133.
54. Lars Steinmetz, PhD et al. The genomic and transcriptomic landscape of a HeLa cell line, *G3: Genes, Genomes and Genetics* March 11: 2013.
55. An extensive expose on informed consent has been discussed earlier in the section on ethics.
56. Eurek Alert *Access to HeLa Cell Genome Data Restored following agreement* (August 2013).
57. See also ClinicalTrials.gov.
58. Nuremberg Code: *Principle #1.*
59. www.*About Canavan Disease | Canavan Foundation*.org.
60. J. Gulcher and K. Stefansson. The Icelandic healthcare database and informed consent, *The New England Journal of Medicine* 342: 2000, 1827–1830; see also Human Genome Organization (HUGO) Ethics Committee. *Statement on Benefit-Sharing*, M. Khoury et al. Commentary: Epidemiology and the continuum from genetic research to genetic testing. *American Journal of Epidemiology* 156: 2002, 297–299.
61. J.A. Todd et al. Robust associations of four new chromosome regions from genome-wide analyses of type 1 diabetes. *Nature Genetics* 39: 2007, 857–864; H. Hakonarson et al. A genome-wide association study identifies KIAA0350 as a type 1 diabetes gene, *Nature* 448: 2007, 591–594; C. Libioulle et al. Novel Crohn disease locus identified by genome-wide association maps to a gene desert on 5p13.1 and modulates expression of PTGER4, *PLoS Genetics* 3: 2007, e58; J. Hampe et al. A genome-wide association scan of nonsynonymous SNPs identifies a susceptibility variant for Crohn disease in *ATG16L1*, *Nature Genetics* 39: 2007, 207–211; International HapMap Consortium. A haplotype map of the human genome, *Nature* 437: 2005, 1299–1320.
62. Nature (editorial). Tribal culture versus genetics, *Nature* 430 489: July 29, 2004; *Havasupai Tribe and the Lawsuit Settlement Genetics;* Paul Rubin. Indian givers, *Phoenix New Times* May 27: 2004; Rex Dalton. When two tribes go to war, 430 in *Nature* 500: July 29, 2004, 500–502.
63. R.A. MacLeod et al. Widespread intraspecies cross-contamination of human tumor cell lines arising at source, *International Journal of Cancer* 83(4): 1999, 555–563 (Cell line: ECV304; V.A. Patel et al. Isolation and

characterization of human thyroid endothelial cells, *American Journal of Physiology: Endocrinology and Metabolism* 284(1): January 2003, E168–E176. PMID: 12388152 Retraction (2009).

64. *WMA Declaration of Geneva—World Medical Association.*
65. L. Burhansstipanov et al. Sample genetic policy language for research conducted with native communities, *Journal of Cancer Education* 20: 2005, 52–57.
66. www.DealSearchOnline.com.
67. Allison Proffitt. deCode publishes largest human genome population study, *Bio-ITWorld* March 25: 2015.
68. Ben Hirschler. Amgen buys Icelandic gene hunter decode for $415 million, *Reuters* December 10: 2012.
69. World Medical Association. Declaration of Helsinki: Ethical principles for medical research involving human subjects, *Journal of Postgraduate Medicine* 48: 2002, 206–208.
70. Xiaomei Zhai. Informed consent in medical research involving human subjects in China. *Fourth Asian Bioethics Conference* (November 22–25, 2002 in Seoul National University, Seoul, South Korea). See also Ding Chuyan. Family members' informed consent to medical treatment for competent patients, *China in China: An International Journal* 8(1): March 2010, 139–150.
71. Presidential Commission on the Study of Bioethical Issues. *Privacy and Progress in Whole Genome Sequencing* (2012): pp 7–8.
72. David Thomasma et al. *Birth to Death: Science and Bioethics*. p 24.
73. T. Dobzhansky, Nothing in biology makes sense except in the light of evolution, *American Biology Teacher* 35(3): March 1973, 125–129.
74. Jürgen May et al. Hemoglobin variants and disease manifestations in severe falciparum malaria, *JAMA* 297(20): May 23, 2007, 2220–2226.
75. Anthony J.F. Griffiths et al. *Modern Genetic Analysis* (W. H. Freeman; New York, 1999).
76. P.W. Corrigan et al. Structural levels of mental illness stigma and discrimination, *Schizophrenia Bulletin* 30: 2004, 481–491.
77. John Davidio et al. *Social Psychology of Stigma* (London, 2000): p 34.
78. Correta M. Jenerette, PhD et al. Health-related stigma in young adults with sickle cell disease, *Journal of the National Medical Association* 102(11): 2010, 1050–1055.
79. E. Goffman, *Stigma: Notes on the Management of Spoiled Identity* (Prentice-Hall; Englewood Cliffs, NJ, 1963).
80. Graham Scambler. Sociology, social structure and health-related stigma, *Psychology, Health & Medicine* 11: August 2006, 288–295.
81. E. Turkheimer, Genetics and human agency: Comment on Dar-Nimrod and Heine. *Psychological Bulletin* 137(5): September 2011, 82.
82. Sophie Lewis. How do obese individuals perceive and respond to the different types of obesity stigma that they encounter in their daily lives? A qualitative study, *Social Science & Medicine* 73: 2011, 1349–1356.
83. E. Goffman, *Stigma* (Spectrum; Englewood Cliffs, NJ, 1963).

84. Michele Easter. "Not all my fault": Genetics, stigma, and personal responsibility for women with eating disorders, *Social Science & Medicine* 75(8): October 2012, 1408–1416.
85. Pamela Sanka et al. What is in a cause? Exploring the relationship between genetic cause and felt stigma, *Genetics in Medicine* 8(1): January 2006, 33–42.
86. Edmund Ramsden. Confronting the stigma of eugenics: Genetics, demography and the problems of population, *Journal of Social Studies of Science* 39(6): December 2009, 853–884 and Edmund Ramsden. Carving up population science: Eugenics, demography and the controversy over the 'biological law' of population growth, *Journal of Social Studies of Sciences* 32(5–6): December 2002, 857–899, *See also* Amy Sue Bix. Experiences and voices of eugenics field-workers: 'Women's work' in biology, *Social Studies of Science* 27(4): August 1997, 625–668.
87. Henry Osborne. *Address at the Second International Congress on Eugenics* (online archives).
88. Ibid.
89. Ibid.
90. Osagie K. Obasogie. Racial alchemy: Bioethics and the skin tone gene, *Science and Society* May 18: 2007.
91. Robert Schwartz. Genetic knowledge: Some legal and ethical questions, eds. David Thomasma et al. *Birth to Death* (Cambridge University Press; London, 1996): pp 33.
92. www.*NIH Genomic Data User Code of Conduct.com*.
93. *Native American DNA: Tribal Belonging and the False Promise of Genetic Science* (University of Minnesota Press; Minneapolis, MN, 2013).
94. Rose Eveleth. Genetic testing and tribal identity: Why many Native Americans have concerns about DNA kits like 23andme, *The Atlantic* January 26: 2015.
95. *Peru's uncontacted tribes threatened by gas project.* http://www.survivalinternational.org/galleries/machu-picchu
96. Keith Wailoo et al. *The Troubled Dream of Genetic Medicine: Ethnicity and Innovation in Tay–Sachs, Cystic Fibrosis, and Sickle Cell Disease* (Johns Hopkins University Press; Baltimore, MD, 2006): 249; A. Sydney. *Halpern Lesser Harms: The Morality of Risk in Medical Research* (University of Chicago Press; Chicago, IL, 2004): pp 232; Nikunj Somia et al. Gene therapy: trials and tribulations, *Nature Reviews Genetics* 1: November 2000, 91–99.
97. Ricki Lewis. Gene therapy's second act, *Scientific American* 310: February 18, 2014, 52–57.
98. Dorothy Romanus et al. Cost-effectiveness of multiplexed predictive biomarker screening in non-small cell lung cancer, *Journal of Thoracic Oncology* 10: 2015, 586–594.
99. Carl E. Schneider. Bioethics in the language of the law, *The Hasting Center Report* July–August: 1994, 16.

3
Direct-to-Consumer Genetic Testing

An introductory comment

An understanding of the etiology of disease has been central to healthcare practice since antiquity. Clinical diagnosis and etiological practices involved some form of interaction with the patient including patients' medical history, observation (of potential symptoms), tests, and other information as needed. Almost all medical traditions have consistently (and rightfully so), incorporated the proper understanding of the causes of a disease into their clinical practice portfolios. At the time of Hippocrates, diseases were superstitiously encapsulated in the cosmological purviews and the actions of the pantheon of the Greek gods and spirits.[1] As a departure from his contemporaries, the Cnidian medical tradition, Hippocrates made an axiomatic shift in the perception of the causes of disease by meticulous observation of the natural world. He suggested that "The body of man has in itself blood, phlegm, yellow bile, and black bile; these make up the nature of the body, and through these he feels pain or enjoys health. Now, he enjoys the perfect health when these elements are duly proportioned to one another in respect to compounding, power and bulk, and when they are perfectly mingled. Pain is felt when one of these elements is in defect or excess, or is isolated in the body without being compounded with all the others." Hippocrates made dexterous efforts to understand the causes of diseases in order to offer the best therapy of the time to ensure that patients regained "balance" in their lives earning him the accolade, the father of etiology. As he once noted, "As to diseases, make a habit of two things—to help, or at least to do no harm." An accurate identification of the causes of diseases was invaluable for the physician to "help" the patient or at least guide him/her not to do any harm. Nevertheless, the very idea to focus on the causes of diseases was revolutionary at the time and current healthcare practices.[2]

Notwithstanding the Hippocratic etiological tradition, Galen or the Galenian etiological traditions also pervaded medical practice for over 1300 years (130–1600). Galen (AD 130) was a Greco-Roman physician, philosopher, a prolific writer. He was highly influenced by the Plato-Aristotelian tradition, the Stoics and his predecessor, Hippocrates. He practiced most of his professional expertise in Rome after many years in the diaspora notably in Asia Minor and Egypt. Following the footsteps of Hippocrates, Galen expatiated on the *Humoral Theory* from which he postulated his etiology as well as the theory of vitalism.[3] The gist of this theory suggested that diseases were caused by the apparent *imbalance* in the constituent parts of the human person. In one of his classics, *On the Art of Healing*, Galen noted[4]:

> When you meet the patient, you study the most important symptoms without forgetting the most trivial. What the most important tell us is corroborated by the others. One generally obtains the major indications in fevers from the pulse and the urine. It is essential to add to these the other signs, as Hippocrates taught, such as those that appear in the face, the posture the patient adopts in bed, the breathing, the nature of the upper and lower excretions … presence or absence of headache … prostration or good spirits in the patient, … [and] the appearance of the body.

Galenian etiological and diagnostics theories and practices and the recommended clinical approach became symbiotic to medicine for over 1300 years until the *Renaissance* when they were deemed antiquated and mostly discarded.[5] While many others emerged, the germ theory especially advocated by Louis Pasteur and later supported by Koch eventually replaced Galenian etiological theories and practices.[6] The *Germ Theory* suggested that germs or microorganisms caused diseases. Thus, the interaction of humans with germs undergirds the presence of diseases. The *Germ Theory* became the de facto etiological theory during the nineteenth and twentieth centuries and was sustained by the development of vaccines and others. Today, etiology has a significant place in the practice of medicine.[7] Lifestyles, the environment, exposure to disease causing microorganisms, and genes are construed to be key indicators of the quality of health of any individual. Increasingly, many technologies continue to be innovated for the proper diagnose of diseases. The advancement in molecular biology and the development of sequencing technologies have highlighted the roles of genetics and its interaction with the environment and lifestyle choices and sometimes profession in understanding and diagnosing of disease in an individual. Historically, most of these theories and technologies were used in clinical settings. However, new computer technologies and molecular diagnostics instrumentations have made it possible to diagnose diseases outside of the clinic or on the blind side of the physician with ease and alacrity. Under certain circumstances, physicians may prescribe genetic tests, partly due to logistical constraints, at other allied health or institutions for diagnostic and prognostic purposes. In particular, the patients have the prerogative and the subtle opportunity to request or even perform their own genetic tests directly from providers.

The phenomenon of gene testing has become a core facet of contemporary biomedicine and clinical research. Genetic-related tests often serve diagnostic purposes in healthcare delivery. The practices have been bolstered, partly, due to the increasing scholarship and understanding of the roles of genetics in human health. A proper diagnosis ultimately leads to proper treatment and prognosis. There are prenatal genetic tests, simple genetic tests for detecting chromosomal abnormalities such as a deletion in the chromosomes, and other tests for genetic mutations. These have led to the emergence and development of several assays and clinical diagnostic tools to determine the clinical implications of basic and sometimes complex genetic aberrations. In a seminal article titled "A Simple Phenylalanine Method for Detecting Phenylketonuria in Large Populations of Newborn Infants in Pediatrics," Guthrie and Susi, A. (1963) presented one of the convincing scientific data and assays for testing a genetically based debilitating disease that affects infants with severe impacts on their mental developments. The simplicity and accuracy of the tests paved the way for *en mass* clinical testing. Sickle cell anemia and Tay-Sachs tests also became popular after the 1970s. Despite the popularity and the demand for these tests, the preferred model has been within clinical settings or as required by healthcare providers or genetic counselors. The past two decades has seen a proliferation of genetic test: the Internet and other social media are inundated with advertisements. Many biotechnology and medical start-ups and labs are offering assortments of genetic tests concurrent with apparent clinical interpretations and diagnoses. A flurry of the genetic advertisement typically asks patients/consumers to submit their saliva, urine, or some form of bodily specimens through the mail ostensibly to the providers within a country or sometimes abroad for the genetic test to be performed. This is known as direct-to-consumer (DTC) genetic tests. According to the NIH, "Direct-to-consumer genetic testing refers to genetic tests that are marketed directly to consumers via television, print advertisements, or the Internet. This form of testing, which is also known as at-home genetic testing, provides access to a person's genetic information without necessarily involving a doctor or insurance company in the process."[8] In addition, a physician may prescribe DTC to his patients.[9] Genetic testing may be ordered for a number of reasons including paternity, traits, ancestral or lineage, clinical or risks for genetic-related diseases, out of curiosity or for indeterminate reasons, which are beyond the scope of the discourse of this book. The DTC genetic tests could be recommended or prescribed by a physician, a genetic counselor, or through the sole initiative of the patient (herein referred to as the consumer/customer). In brief, Direct-to-consumer genetic testing kits are marketed to people who aren't necessarily ill or at high risk for a disease, but who may be just curious or concerned about their risk for different disorders. Some of these tests require a physician's prescription, but many are sold directly to consumers on the Internet. The commercial tests examine a small number of the more than 20,000 genes in the human body and, in theory, predict your risk for conditions such as heart disease, colon cancer, and Alzheimer's disease; determine disease carrier status for pregnancy planning; and identify genetic variants that increase or decrease your ability to metabolize alcohol and certain drugs. Many also offer ancestry tracking—identifying clusters of gene variations that are often inherited by a group of people with a common origin.[10]

The DTC genetic tests are based on business models and as such targets the patient or customer. The business modus of operandi of most DTC genetic tests are typically shrouded in such seeming candor, anonymity that precludes a patient's health provider (physician). These features of DTC genetic tests have generated some rankles within the clinical and scientific communities, healthcare insurance, regulatory, and other stakeholders and interest groups. It seems to delineate the role of the physician and healthcare providers from the Hippocratic medical tradition of holistic care for patient including diagnostic services. As Hippocrates once noted, "It is more important to know what sort of person has a disease than what sort of a disease a person has." DTC genetic tests service potentially reduces clinical diagnostic roles of the physician including a holistic understanding of the patients and other potential causes of diseases and health to a vestigial position. Some proponents argue that this model of providing genetic test directly to the patient signals the new era or at least the threshold for personalized medicine especially within the corpus of the Hippocratic traditions' model of *house-calls*! Of course, the direct provision of genetic tests to patients (on the premise that the tests are accurate), changes the dynamics of healthcare partly into the hands of patients who proactively engage in the process of clinical health care for themselves. Some bioethicists and regulatory bodies have raised concerns about the DTC model of genetic tests partly because there is hardly any regulatory oversight regarding the tests, the labs, and the *experts* conducting the analysis and interpretation of the results. Several analyses of the pros and the cons of the DTC have been adduced. I intend a seismic shift in terms of methodology. In this chapter, I offer a *dialectic analysis* of the DTC genetic tests and the potential import for personalized care. I will examine the theses for and against DTC genetic testing, cognizant of the new impetus toward precision medicine and personalized health care.

DTC genetic tests: Dialectic analysis-implications for personalized medicine

One of the core arguments in support of the DTC model of genetic tests is that it purportedly offers *faster* diagnostic services to patients or consumers; perhaps in comparison with testing in clinical settings due to established bureaucracies. For instance, 23andMe (3–4 weeks), The Genographic Project or Geno 2.0 (8 weeks), Ancestry DNA (4–6 weeks). DTC also offer varying Internet-based supportive services: from the date of receipt of biospecimen, lab analysis and tests, interpretation, biobanking of biospecimen, and access to test results in real time on the Internet. These tests could unduly delay in clinical settings especially where the clinical facility does not offer such tests or do not have the lab and equipment needed and rather outsources them. In clinical settings, a physician may need an approval from insurance carriers while DTC genetic test does not. In addition, patients may be required to make additional appointment to meet with their physicians and genetic counselors in order to know their health status and genetic test results, whereas in DTC model, the test procedure and results are communicated expeditiously to the consumer. Proper and timely diagnosis of certain genetic proclivities or diseases is critical in offering therapeutic interventions or counseling.[11] For instance, timely identification and diagnosis of the BRCA gene could lead to a faster clinical strategy in terms of therapeutics and care for the patient.[12] Furthermore, a swift genetic diagnosis offers an arsenal of clinical information to the patient, healthcare providers, insurance companies, and families. Clinicians could rule out suspicions of diseases and narrow symptoms to other potential causes of diseases especially if the condition seems clinically indeterminate. Also, diagnostic information may be helpful for either long- or short-term purposes depending on the outcome of the genetic tests. It does make clinical sense for patients to explore expeditious testing from other clinics that offer DTC in order to access care if the results are positive.[13]

The second most cited thesis for DTC is genetic tests and are *accessible* to consumers/patients.[14] Obviously, the business model considers the "patient" as a "consumer." As such, DTC genetic testers offer excellent customer care from the collection of the biospecimen, by making it easy for the "customer" anonymously and without prior appointment. The customer simply submits the biospecimen in a kit and indicates the kinds of genetic test he/she wants and pays accordingly. In effect, the customer-centered approach seems to ensure that patients are in-charge of the tests. It is akin to the *Uber effect* where the customer is the ultimate focus who directs the pace and the extent to which the genetic test is performed. They also offer additional and important services such as creating unique portfolios of identifications such as numbers to customers so that they could log into their respective websites and follow the test procedures; sometimes in real-time fashion contrasting actual clinical tests where similar tests are usually "protected" or "guided" by some established clinical operational norms. In brief, bureaucratic procedures associated with genetic tests in clinical settings are short-circuited; apparently by the ease, seeming anonymity, and accessibility of the DTC genetic tests by the patient/customer. Furthermore, in an era of globalization, an easy access to DTC could obliterate the burden of waiting many months or years to undertake some genetic tests in clinically underresourced communities who might lack such facilities and equipment for genetic tests. The proliferation of the DTC genetic tests could also lead to a reduction in the fees associated with the tests. This is possible if there is a high demand and many clinical facilities become easily available based on the economic principle of supply–demand.[15]

The third thesis for DTC genetic testing is described as *diagnostic autonomy* and vacillates on the notion that patients take the *initiative* for genetic testing either out of curiosity or for clinical purposes and out of their own volitions.[16] It is common practice for consumers to order DTC for social, legal, or even religious reasons. Others also take the initiative to request for DTC genetic testing in order to identify and compile their genetic or ancestral links, obtain ethnographic information and genetic markers. It seems some of the most successful DTC corporations like Geno. 2.0 started in this way and eventually added some of the clinical components in response to market demands and changing trends. Regardless of the outcomes of the DTC genetic tests, patients or consumer's intrinsic capacity and initiatives is seen as a leap in the practice of medicine and signal the shift in perspectives toward personalized medicine. In addition, diagnostic autonomy is a demonstration of individual's proactive decision toward preventative care rather than curative care. It seems most of the models of medicine has been curative care in which patients are diagnosed in clinical situations during a visit to the healthcare provider or during an ER visit. Thus, there is a shift in power dynamics and *decisional capacity* toward the patient or consumer. This is because the patient is able to determine and make his/her own decision regarding the DTC genetic test, which in essence reflects a shift from the residues of clinical paternalism. Patients do not necessarily need the fiduciary prescription of healthcare providers (though physicians could and do prescribe DTC) in as far as they are above the legal age and able to consent to the test.[17]

DTC empowers patients/consumers to make clinical or quasi-clinical decisions that serve their personal interests and the rights to their bodily autonomy with little or no undue external influence. DTC customers seem to have a swath of clinical information ahead of their respective healthcare providers and will invariably be in a position to "personalize treatment" or ask for precise therapeutic intervention that suits them through their own initiative. For example, if a sickle cell patient knows her status through DTC, she could be in a better position to prepare for any medical emergency associated with it by making certain changes accordingly. In addition, in the midst of medical emergencies and should the patient become incapacitated, a preexisting DTC genetic test result could potentially be useful and timely for healthcare providers to offer clinical care and insights into their health status and genetic predispositions toward illness.

Fifth, short- and long-term *quality of life planning* are essential to almost every one. People meticulously plan every aspect of their lives such as education, career, healthcare, and types of leisure they want, retirement, old age, and even have elaborate plans for their funerals in the event of death. Most of these plans may be contingent on the quality of their health status or genetic risks for diseases. DTC genetic tests and clinical profile will serve as a template for patient to make competent and informed decisions and potential lifestyles changes at the personal level or with their respective families. In addition, DTC genetic test results may be applicable in formulating public health campaigns and education based on the prevalence of certain debilitating genetic markers in the population, especially if this could be of epidemiological imports.[18] There is no doubt that people are living longer compared to previous generations. Vaccinations, increase access to clean water, cleaner environments, access to health care, decrease in infant mortality and fecundity, increase in wealth and other socioeconomic and clinical indices are favorable factors in increasing the average age of the global population. It is equally important to note that despite these factors, there are still populations that have lower age range in terms of longevity. The longer we live, the more resources and services we dissipate naturally including clinical and gerontology-related care. Quality DTC genetic tests can help patients develop specific road maps at various epochs of their ageing process. Some genetic-based diseases such as Crohn's, Tay-Sachs, sickle cell, and genetic-related vision problem, neurodegenerative like Alzheimer's and others are incapacitating requiring extensive sociomedical care. A routine DTC genetic test may help the patient plan for the future especially during old age in order to have a reasonable quality of life or gerontology-based decisions. Living longer does not necessarily confer quality of life and health per se but knowing these possibilities is crucial in preparing for any predicament.[19]

As a consequence of the above, DTC genetic tests can also inform patients in order to purchase appropriate health, life, and financial insurance to suit their current health or impending health status especially where there is no universal healthcare coverage/single-payer system. A genetically informed customer could optimize his/her insurance benefits by buying the best possible plan or even additional plan(s) to insure proper coverage and care. This is even possible in the short term where certain insurers might consider some genetic diseases or risks as preexisting conditions. The patient might disclose these risks or not depending on the legal or policy governing these issues in context. Long-term policy planning are major decisions for patients who might have disposition toward chronic diseases such as Parkinson's, Alzheimer's, and cystic fibrosis. Some of these have significant price tags in terms of care and overall burden on national or global economic health expenditures. Several studies and projections have shown that patients with these diseases could have appreciable quality of life if they have access to quality clinical and socioeconomic care. In a recent study conducted by Mudivanselage and others, they noted, "PD [Parkinson's disease] is associated with significant costs to individuals and to society. Costs escalated with disease severity suggesting that the burden to society is likely to grow with the increasing disease prevalence that is associated with population ageing."[20] Currently, patients with Parkinson's disease spend about $32,556 Australian dollars per year for their health care while the social burden is estimated at $45,000. In the United States, the annual economic burden is projected at $22,800 per year per patient.[21] Thus, it is pragmatically feasible and advantageous for an individual to know his/her genetic profiles in order to make reasonable financial projections and plans for the current and the future. In cultures where the onus of health care and social services are insurance-based rather than familial responsibilities, DTC genetic tests seen as indispensable tool for customers to procure specific or personalized insurance premiums to meet their specific needs. Personalized health care requires personalized insurance package and protection. As in the words of the FDA, genetic tests are "intended to inform users of lifestyle choices and/or encourage conversations with a healthcare professional." In other words, genetic tests are not ends in themselves but rather help patients to have further clinical consultations and discussions with healthcare providers in order to make choices or sometimes no choices at all. Several epidemiological and oncological studies and data suggest that African-American males have high incidence of prostate cancer compared to males in other populations. As with many cancers, an earlier identification helps in proper treatment and thus enhances the rate of survivability of patients. DTC genetic tests and regular tests including clinical evaluation of patients can save lives and curtail the development of prostate and other genetic-induced cancers in this population. In brief, access to reliable and clinically valid DTC genetic tests makes pragmatic and socioeconomic sense.

Seventh, quality public health campaigns and education are at the forefront of health care. Genome-wide studies coupled with quality genetic tests within population holds sway for the development of robust personalized public health campaigns and education in ameliorating the emergence of epidemics and diseases. Iceland is cited by scholars as having one of the earliest models of genetic-based public health education program. deCode has pioneered one of the most successful population-based genomic sequencing in the world. The company and its affiliates have collected and analyzed many biospecimens, and have several genetic tests and as I have noted above, sequenced the genes of most of the population in Iceland and have created an extensive database. Iceland is unique because of its homogenous gene pools due to the possibility of its population from a homologous ancestry and centuries of isolation from the rest of the world. In addition, the population has kept extensive ancestry records making it easier to track generations. Because of these factors, geneticists are able to decipher some specific variants in the genes of sequenced genomes of the population responsible for diseases or phenotypic manifestations. In fact, there have been consistent efforts in encouraging Icelanders to sequence their genomes. A DTC genetic tests (which appears popular in Iceland), is essential for designing public- or population-based campaigns and education for individuals to make personal decisions about their health and families. DTC genetic tests if popularized strategically across the world can serve as important tool to help in designing, disseminating, and educating the public about their health in terms of the genetic markers and risks. For example, researchers believe 0.08% of the female population in Iceland have the mutation for the BRCA 2 genes that causes breast cancer with the clinical probability of carriers developing cancer to be over 80%! Another example is macular corneal dystrophy—an inherited eye disease. Compared to other population samples, Iceland has one of the highest incidences of this eye disease.[22] As some scholars in a recent study noted, "The fact that the common ancestors of the parents of persons with macular corneal dystrophy were not found until in the 18th century, together with the available information of the geographical whereabouts of the families for more than two centuries may permit us to assume that heterozygous carriers for corneal dystrophy were present in Iceland early in the 18th century."[23] In simple terms, the ophthalmological condition is present in certain populations who are carriers of the gene. Thus, a public health-based campaign may target the population noted to be carriers of these genes to take appropriate precaution for clinical interventions, genetic counseling, and/or other decisions to manage the inheritable eye condition.[24]

Genetic tests are no doubt important in the shift toward personalized medicine, pharmacokinetics, and pharmacogenomics. DTC genetic tests are invaluable in clinical research and in the development of pharmaceuticals. Individuals have different metabolic pathways to the extent that patient reacts differently with biopharmaceuticals. These include tolerance, metabolism, and rates of absorption and excretions of biopharmaceuticals. In a recent interview with CNN, Stefansson suggested that "...roughly half of Amgen's current research projects are influenced directly by genetics, and of those, at least 90% stems from the work happening at deCODE."[25] Other biopharmaceutical corporations are also conducting significant research into genetics and its association with diseases in order to design personalized pharmacological solutions. As indicated in the preceding chapters of this book, the completion of the human genome in particular is pushing the frontiers of science and the practice of medicine toward precision or personalized medicine. Availability of data on genes, mutations, and disease causing genes offers a plethora of hope for charting these paths and paradigm shifts in medicine and allied health sciences. DTC offers participatory roles for consumers to be part of the paradigm shifts. As significant numbers of the population order for the DTC genetic tests, either covertly or overtly, they participate in the "health delivery" system by their active initiative in DTC genetic tests. It is a gratifying situation to see many people out of curiosity and other medical intents striving to know their health status actively engaged in these diagnostic enterprises. In other words, individuals are taking "charge" of their health and as I have expatiated above, the power dynamics relationship between patients–healthcare providers seems to be having some axiomatic shifts. The entire procedural perspective of DTC genetic tests is also an educational process for patients. Irrespective of their academic, professional background, or social status or age, patients participating in any form of DTC genetic tests will get some scientific education or at least in pedagogical terms undergo some conceptual change about themselves in relationship to the larger population about the nature of genetic science. The collection of the biospecimen itself are often laborious and meticulous, and scientific processes and customers will be required to follow a specific protocol sent to them in order to ensure the scientific integrity of the tests. They also have the opportunity to read important genetic information and data, thus stretching and strengthening their scientific literacy and perhaps better appreciation of the practice of medicine and public health. A scientifically literate patient may be in a better position to make "informed" decision regarding his/her health and the DTC genetic tests seem to encourage and sustain this.

Despite the above factors in making the case for DTC genetic tests, some experts within and outside of the healthcare professions have raised genuine concerns as with new technologies and ventures. Some have identified some limitations in the DTC genetic tests business models, the interpretation (and potential misinterpretations of tests results), the secrecy surrounding the testing labs, lack of regulatory oversight and quality control, and the potential for discrimination and abuse. These concerns are diametrically in contradistinction to the theses for DTC genetic tests and I intend discussing some of these below.

Some opponents of DTC genetic tests point to parallels to some of the challenges of the Theranos scandal. Theranos, a company founded by Elizabeth Holmes, offered many blood-based tests on the *DTC* model. Theranos claimed among others that they have the scientific resources and capabilities to draw relatively small amount of blood from patients compared to clinical labs and were able to perform an array of tests. The *fingerprick blood* methods became the marketing garniture of Theranos and the company even received legislative backing in Arizona (the Clinical Labs Improvement Amendments) to the extent that patients could directly request the tests without a clinicians' prescription. However, nearly two years after reports emerged challenging the scientific accuracies of Theranos services, the company has been besieged with several reprimands from the FDA and others cascading in the shutdown of some of the major testing operations with vendors such as Walgreen truncating their contractual obligation by refusing to sell the fingerpicks blood tests products. One of the most prestigious scientific journals, *Nature*, in a tersely worded editorial noted:

> Time and again, new health-care firms are forced to realize that it helps no one to be secretive with data. Even if it turns out that the Theranos technology does not work as well as advertised, the company would hardly be the first to find itself in that situation. Releasing more information earlier might have forced Theranos to confront shortcomings. Instead, it finds itself trying to recover from a regulatory and public-relations hole. This is not an insurmountable situation, as 23andMe knows. The challenge now is for Theranos to show us the data.[26]

DTC genetic tests akin to the Theranos customer-based business model have become a source of concern for clinicians, researchers, health policy experts, and health insurance carriers. The proliferation of DTC and other genetic tests has become unprecedented in the past two decades but have remained largely unregulated at the national or international levels.[27] However, efforts and initiatives at the professional levels, some international guidelines from the World Health Organization and other organizations, the proliferation remains seemingly unchecked and surreptitious. As the *Nature* editorial keenly noted above, DTC tests are markedly secret and they barely share data or information regarding their procedures and results. What is the reason for a company to gamble in keeping the test procedure and results of a patient in dire secrecy? Indeed, as the Theranos case and the 23andMe seem to point out, there is an increasing quest for public disclosures and scrutiny for the good of the companies and the patients as well. A cursory look at most of the DTC genetic tests corporations, however, raises concerns as similar initial mistakes made by Theranos and 23andMe seem to be recurring. Of course, 23andMe seem to have learned from these mistakes and currently have FDA imprimatur to continue with some of their DTC genetic tests. Would the avalanche of other DTC corporations emulate from 23andMe and accedes to proper public and scientific disclosures of their products including the procedures in testing for their clients?[28]

One of the convincing arguments against DTC genetic test is the lack of regulatory overture of the operations of the corporations.[29] Genetic tests such as DNA fingerprints, actual fingerprints, and paternity tests for forensic and legal purposes are highly regulated globally. Their labs including location, equipment, and personnel (including their education, training, and suitability) are subject to regulation to ensure scientific accuracy and integrity of the test results.[30] However, DTC genetic tests are almost surreptitious in terms of the kinds of equipment they use, protocols, whether they use assays or actual clinical test data, personnel, interpretation, and validity of the results among others. One of the hallmarks of good science or scientific enterprise is peer reviews, disclosure of the scientific methods for others, as well as the possibility of replication of concepts and protocols by an independent and competent scientific entity vested with public interest. In other words, a *scientific claim* or proposition must be publicly demonstrable for it to be part of the corpus of public knowledge to an extent protected by trade secrets and intellectual property legal provisions. Due to historic precedents of some tragedies in the annals of science, some regulatory bodies such as the US Food and Drug Administration have provided consistent leadership in regulating science over a century now (as detailed in subsequent chapters of this book). Regulatory oversight has proven to guarantee public safety and has prevented the marketing of pseudo-scientific products and ideas. For instance, almost all (bio) pharmaceuticals and most medical devices go through rigorous approval processes. Despite the wanton proliferations of DTC genetic tests, suffice it to say, there is lack of regulatory aperture and it appears the modus operandi of most of the DTC corporations are not even known in the scientific communities for the validity of their products. It does not mean, however, that the DTC genetic tests and their interpretations may not or may be scientifically accurate and clinically valid. Rather, the lack of regulatory overture to validate and authenticate the DTC tests has instead created a cloud of uncertainty within a segment of the population. Some healthcare providers prescribe the DTC genetic tests to some of their patients while others have concerns due to lack of regulation of the products. It is refreshing to point out that the FDA has recently given a federal imprimatur to 23andMe for their DTC genetic tests after many setbacks *in tandem* with the Theranos scandal. While many see this to be a step in the right direction, others feel the current approval for 23andMe is perhaps a knee jerk regulatory intervention rather than a holistic detailed regulatory oversight similar to the drug approval process. Should all DTC genetic tests have a single bioinformatics depository for test results akin to CODIS, the FBI database in ensuring quality assurance?

The practice of quality control and assurance are consistent with quality products and services. International regulatory bodies such as Health Canada, European Medical Agency, the US Food and Drugs Administration, and others have detailed guidelines on quality assurance for the regulation of the biopharmaceutical and biomedical devices and products. Corporations adapt these quality assurance guidelines and protocols and document them throughout their application for approval for their products. On the contrary, DTC genetic test are dependent on their respective internal quality control mechanisms in terms of the kinds of tests and services they offer. Quality control and standardization of equipment may ensure uniformity in test results. Whereas in biologics, corporations adhere to a common or highly similar procedure quality control protocols, DTC genetic tests do not seem to have a universally defined and verifiable standard. There have been few reported cases of discrepancies in DTC genetic test results and many have wondered if some form of standardization and quality control mechanisms could avert such discrepancies. One of the most cited were triplets believed to be maternal and, therefore, genetically identical—the Mynard sisters. They had their genetic tests through the DTC model with 23andMe AncestryDNA and Family Tree DNA. According to an extensive news coverage by the *Insider* where they offered the detailed results (later corroborated by the DTC genetic tests corporations), several discrepancies in the results were discovered to their chagrin. According to the report, "Nicole was 11 percent French and German but Erica was 22.3 percent. Their sister Jaclyn was in the middle at 18 percent" in addition to other variations in the tests' results depending on which of the three entities was being tested. Other results were identical even from the three different companies. While it is common for identical twins and even triplets to have some genetic variations or SNPs, the variations in the DTC test results are stunning, particularly if these were for actual clinical use. In the wake of these, Family Tree DNA offered to change their testing/method *algorithm*![31] In other words, they were going to improve upon their quality control methods to ensure accuracy. In brief, the apparent discrepancies are sufficient to continue the conversation and touting the case for standardization and quality control measures as *sine qua non* for DTC genetic testing and the roles they may have in personalized medicine.[32]

One of the main theses against DTC genetic is the plausibility of misuse for self-diagnosis. The substance of the argument is that DTC genetic tests are entirely or mostly self-initiated and voluntary in nature.[33] In addition, the genetic tests results are usually communicated to customers with or without genetic counsels or competent medical intermediaries. As such, it is possible for patients to self-prescribe and treat if the DTC results determine the presence of a disease risks or biomarkers for a disease. While self-prescription and treatment may be very difficult in most countries, it may not be an unreasonable practice in some parts of the world. There is always a danger to self-treat as the patients do so at their own risks and the risks could be compounded by the fact that DTC genetic tests interpretations and results are not necessarily and clinically valid. In the case of BRCA 2 gene mutations, self-treatment could be fatal as well as any delay in seeking for highly specialized therapeutic intervention including chemotherapy and current clinical practices, which is best done by a licensed clinician and at approved healthcare facilities. Furthermore, a DTC genetic test result is sufficient for patients to embark on medical tourism, especially gene therapies and other pseudogenetic treatments that might not be offered in their own countries or healthcare systems. With the stem cell therapy as a point of departure, Alta Charo insightfully observed in a recent piece in the *New England Journal of Medicine*:

> In 2011, football quarterback Peyton Manning went on the road to seek out stem-cell "treatment" for his neck. He wasn't alone: many high-profile athletes and desperate (but less famous) patients left the

> United States seeking interventions available in countries with less rigorous regulation. They didn't necessarily know what kind of cells they were getting, whether there was any evidence the intervention worked, or whether anyone understood the risks they were taking. So why did they do it?[34]

She speculated that the human drive (albeit a fallacious one) toward the new, *argumentum ad novitatem*, may partly explain why patients participate in medical tourism irrespective of the risks. DTC genetic tests and the therapeutic potentials of gene editing tools present a new frontier for medical tourism across the globe for vulnerable and desperate patients. Indeed, given the stories about amazing potential and early breakthroughs in laboratory and animal models, gene editing may trigger another wave of medical tourism. We should take steps now to guard against future gene-editing tourism by developing professional norms, fostering collaboration among national regulatory bodies, partnering with patient-advocacy groups to develop accurate, credible information sources, and working to devise responsible research protocols and patient-monitoring measures.[35]

The question of preexisting condition is like a *gadfly* in the forefront of medical practice and health insurance purchase. According to the US Department of Health and Human Services, a preexisting condition is "…a health problem you had before the date that new health coverage starts." Prior to the Affordable Care Act (ACA), some chronic medical conditions such as diabetes, some cancers, some mental disorders, epilepsy, injuries, and pregnancy constituted preexisting medical conditions. Patients who had them prior to enrolling or buying a health insurance premium, paid more or sometimes lost their health coverage completely. In fact, most health insurance companies and subsidiaries had the legal discretion and authority to preclude patients from coverage due to the usual argument of higher risk costs associated with treatment or managing these conditions. However, the ACA has obliterated this approach and now these hitherto preexisting medical conditions are covered by most health insurance, though the health insurance corporations still have the discretion to increase premiums for patients in these categories. Despite the changes in the law regarding preexisting conditions, genetic illness remains a highly contentious issue especially in the global healthcare delivery system. The new shifts toward personalized medicine may heighten these concerns partly because DTC genetic tests can and do identify many adverse biomarkers within the human genomes. However, SNPs and associated genetic preexisting conditions do not correlate or mean that an individual will manifest a disease (though in some cases: BRCA 2 has a very high possibility). Furthermore, there is no such thing as a clean human genome. Owing to the evolutionary trajectories of human emergence, adaptation, diseases, and other environmental stressors, genome-wide population studies so far have some polymorphisms or specific changes in the genomes. These have made individuals unique carriers of genetic aberrations. Some of these aberrations are benign; some potentially leads to diseases, while many others remain amorphous or undetermined. In brief, our genes define our physiological, neurological, anatomical structure and make up, coupled with our interactions in the micro and macro environment. Consequently, a preexisting medical condition inclusive of genetic aberrations remains highly contentious. Nevertheless, the existential reality is that there is a phobia that genetically induced medical conditions may attract significant treatment or management and as a result, insurers (given the opportunity), will charge higher premiums in exchange for health insurance coverage. The costs associated with healthcare and social responsibilities for AL patients appear high. Nevertheless, such demonstrable expenditures may constitute a justifiable reason to increase premiums in terms of preexisting medical conditions. Axiomatically, the proliferation and easy access to DTC genetic tests will likely increase the number of patients with "preexisting genetic conditions" as the genetic tests may highlight those with markers for many of the diseases mentioned above erroneously deemed as preexisting medication conditions. In view of the discussions above, should an individual initiating his or her own DTC genetic tests legally disclose a known genetic medical condition to his/her health insurer? Should insurers increase their premiums due to the risks and possibility of long-term care and "burden" on the insurance pool? In some situations, nondisclosures of preexisting and genetic aberrations are deemed fraudulent. What about discrimination even though the Genetic Information Nondiscrimination Act of 2008 (GINA) is still in place in the United States? GINA was promulgated to specifically insulate and protect patients from any form of discrimination due to their genetic disposition and associated biomarkers for risks, family ancestry, genetic services, genetic research, and others.

Preexisting genetic medical conditions have implication in the employment arena too. On a positive note, genetic tests and interpretation of results are very informational but may also help an employer accommodate an employees' predisposition toward a risky work type and environment. Some genetically identifiable diseases need constant care and exclusion from some types of work in order to prevent injuries and optimize productivity. For example, sickle cell patients generally have good health if they know their health status and intentionally plan and adjust to the symptoms of the disease. During sickle cell "crises," carriers of the mutation generally exhibit symptoms of physical weakness; chest, lower back, stomach pains, and dizziness. These mean patients are likely to miss work or apparently have lower productivity during crises. However, if disclosed, other work not involving physical activities may offer excellent opportunities for sickle cell patients to succeed and advance professionally. Also, carriers of the Huntington's disease gene have severe limitations during their life time due to the degenerative nature of the disease. HD patients' lives are often exacerbated since there is no cure per se of the actual mutation—most carriers of the gene manifest symptoms at an early stage in their lives and have generally shorter life spans. As a result, they need quality clinical management and care from their families. Considering our current seemingly competitive work culture, employment for HD patients could pose some challenges. Surprisingly, some employers seem to be equally proactive in identifying employees with genetic markers as at risk to their work environment. Recently, the

US Equal Employment Opportunity Commission (EEOC) settled a genetic discriminatory case involving Burlington Northern Santa Fe (BNF) Railway and their workers. BNF Railway authorities have repeatedly documented evidence of a high incidence of work-related injuries involving employees. BNF requested worker's unions nationwide to subject any employee reporting "work-related carpal tunnel syndrome" to genetic tests. Studies assert that *Hereditary Neuropathy with Liability to Pressure Palsies* (HNLP) causes some work-related carpal tunnel syndrome. BNF theorized that most of the reported injuries were probably preexisting or genetically predetermined conditions rather than work related.[36] If these assumptions are clinically authenticated genetic testing, it may lessen their legal obligation and liability to the affected employees. Therefore, BNF started genetic tests program for its employees surreptitiously without their knowledge, expressed, and full consent. One of the employees of BNF Railways who objected to the genetic tests was threatened with termination. Accordingly, the workers sued and the EEOC eventually arbitrated the case out of court with an injunction against BNF. The EEOC also issued an "Agreed Order" with the following legally binding terms and conditions[37]:

a. BNSF shall not directly or indirectly require its employees to submit blood for genetic tests.
b. BNSF shall not analyze any blood previously obtained.
c. BNSF shall not evaluate, analyze or consider any gene test analysis previously performed on any of its employees.
d. BNSF shall not retaliate or threaten to take any adverse action against any person who opposed the genetic testing or who participated in EEOC's proceedings.

In another related case, some employees who had worked (including some current) at the Lawrence Berkeley Laboratory belonging to the Department of Energy at California, under the aegis of the University of California reported being discriminated on the basis of their genetic profiles. Some of the employees of color reported that during employment screening, a third party contracted by their employer was secretly testing their biospecimen for sickle cell traits, pregnancy, and checking their syphilis status between 1972 and 1995 without their consent. They filed a lawsuit in 1995. They challenged the legality and circumstance of the tests. They averred that such results were probably used, and will likely be used to intentionally discriminate against them thus obliterating their professional and economic opportunities. In addition, the test could also reveal their familial health and genetic status. The lab refuted the allegations in the court of law and the US District Court in California accordingly dismissed the case in 1996.[38] However, the initial decision was reversed in the US Court of Appeal in 1998 in favor of plaintiff. In the opinion of Judge Stephen Reinhnardt, "The conditions tested for were aspects of one's health in which one enjoys the highest expectations of privacy."[39] The lab agreed to an out of court settlement with the parties. Despite the hefty settlement, defendant denies any discriminatory practices. These two examples have become *locus classici* of the high-profile genetic-based discriminatory cases in the United States.

As the legal dictum suggests, he who alleges must proof but genetic cases may be difficult to substantiate especially when a third party have access to the biospecimen of employees and could potentially biobank it for a period of time. There is a growing local and international consensus for specific policy guidelines and strengthening local norms on genetic testing and the possibility of abuse. Having a genetic mutation or marker is not a disease and appropriate measures including policies should insulate patients from the tentacles of discrimination in the workplace and others. As Otlowski et al. noted in a recent study,

> Research which validates the claim that genetic discrimination is occurring has been limited, both in scope and design. There has, as yet, been no comprehensive co-ordinated empirical research about the nature and extent of genetic discrimination across countries where genetic services are highly developed. More significantly, the studies undertaken to date rely predominantly on unverified and in many instances, anonymous accounts of individuals' subjective impressions of whether they received inequitable treatment from third parties such as employers or insurers. Although new initiatives are now being undertaken within the insurance industry in this regard,[9] there has also been a general absence of systematic documentation research into current third party policies and practices, by which responses to issues associated with the genetic profiles of individuals are determined.[40]

Rather, opportunities for proper clinical monitoring, care, treatment (if possible), availability of social services and access to quality education, employment opportunities, and accommodation are essential for managing potential or adverse genetic profiles of individuals. There is a concern that a lack of clear legislative and legal framework on genetic testing could create an underclass of people lurking under burdens inadvertently due to their genetic proclivities and preexisting medical conditions. As Vincent noted in the movie, GATTACA, "I belonged to a new underclass, no longer determined by social status or the color of your skin. No, we now have discrimination down to a science." But preexisting medical conditions are an essential part of who we are as existential beings. Every person has a genetic aberration and therefore a preexisting medical condition. We should embrace genetic pluralism in view of the inalienable common human bond we share as intelligent beings with the capacity in caring for each other are profound and timely.

Another thesis against DTC is on biobanking. Technically, a biobank is defined as "an organized collection of human biological material and associated information stored for one or more research purposes" such as genetic testing and array of others.[41] These include the collection and storage of tissue and organ samples, blood, eggs and sperms, cadavers, cancer cells, and whole genomes. The collection and preservation of human specimens (biobanking) is not new. For example, even though the Egyptians mummified the Pharaoh's for sociocultural reasons, this practice has preserved/

biobanked valuable biological information for researchers and scientists. In a recent edition of the *Nature* journal, it was reported that scientists were able to sequence and analyzed the DNA of some Egyptian mummies, believed to be of the Pharaohs, apparently revealing the mummies' family relationships as well as their afflictions, such as tuberculosis and malaria providing unprecedented insight into the lives and health of ancient Egyptians and is ushering in a new era of "molecular Egyptology."[42] Genetic testing is applied in forensic anthropology and forensic paleontological studies to ascertain some historic assertions and controversies. Scientists have also used DNA and genomic tests to understand some epidemiological issues in the past and an array of others. In modern times, many scientists out of their own volition collect biospecimens such as tissues, blood, cells of various kinds, and cadavers for diagnostic, etiological, pathological, and research purposes. Biobanks serve as depositories for biospecimens for DTC genetic testing. DTC begins with collection of biospecimens, which are stored and later used for testing. However, it seemed during the1990s that a new wave of scientific and research interests might have galvanized the scientific and research community to collect diverse samples of biospecimens. As one scientist, Dr. Carolyn Compton of the NCI's Office of Biorepositories and Biospecimen Research poignantly noted, "Biobanks will transform the way we see disease developing."[43] For "Ten years after the human genome project, the potential for personalized medicine lies not in a single genome but in many…" Hence the "collections of biospecimens that are accompanied by data on medical history, behavior and health outcomes are crucial to this task" among other things.[44]

As briefly noted, the Parliament of Iceland promulgated a law in 1998, ostensibly for the collection of biospecimens and the national storage of data accrued from these collections. In 1999, the US National Bioethics Commission issued some policy guidelines and gave extant recommendations on the creation and the storage of biological specimens. This led to the creation of the Office of Biorepositories and Biospecimen Research under the auspices of the National Cancer Institute. The European Union also promulgated its own version of policies geared toward the proper handling of human biospecimens in 2006. These international and intranational efforts resulted in a tsunami of collections of biospecimens throughout the world. It is estimated that over 350 million human specimens have been collected and *biobanked* in various labs and facilities throughout the United States alone.[45] The significance of biobanking is inherently evident. As already noted in the introductory paragraph, biobanking has many implications. If properly regulated, biobanking will allow for ample collaborations and corroborations in research, diagnosis, and the development of novel therapies for the benefit of humanity.[46] In addition, preservation of biospecimens is economically sound and cost effective for scientists; for example, HeLa cells are a multimillion enterprise for researchers who have preserved these cell lines since 1951 and needless to say that various studies in these cell lines have generated copious and valuable data on cancer. Biobanking also makes valuable biological specimens such as organs, tissues readily available and accessible, and prevents extinction. Biobanking is also an invaluable source of materials and information for the ever-emerging fields of recombinant biotechnology/bioengineering, organ, transplantation, virology, immunology, pharmacogenomics, pharmacology, forensic science, among others. However, DTC-operated biobanks are subject to internal regulatory oversights of the respective corporations. Biospecimen sent by consumers for DTC genetic testing remains the property of the testing entity as discussed in the preceding chapter on confidentiality and privacy. Specimen from consumers is partially anonymous because the veracity and authenticity of the consumer is based on the presumption that they are real, whereas in clinical contexts the patients are mostly physically present during the collection of the biospecimen. Can these pose both operational and integrity challenges? Should biospecimen for the DTC genetic testing be clinically verified locally or at least have fingerprints or basic DNA tests locally to accompany the biospecimens? Responses to these operational but significant questions remain an open debate in this early stage of personalized medicine. Furthermore, some ethical questions oscillate around *privacy, autonomy,* and *confidentiality.* As one prominent physician pointed out, "Having all of your DNA out there where organizations or governmental institutions have access to it makes people nervous."[47] Many people have raised these questions because genetic materials could and have been extrapolated from some specimens and made available to the public or unauthorized users. This will constitute a breach of the ethical principles of privacy and confidentiality. The concern here is that if the donors personal biospecimens data ends up in the hands of health insurance companies, there is the high proclivity and propensity of being denied adequate coverage if there are potential biomarkers for certain diseases such as cancer, diabetes, among others.[48]

Crime Science Investigation (CSI), CNNs Forensics, and Forensic Files are among some of the most popular documentary programs across the segment of society. CSI seems to extemporize forensic evidence such as DNA in crime scenes. Research suggests that some jurors are influenced by these shows especially in making juridical decisions and adjudication in the court of law. Scholars have described this phenomenon as the *CSI effect*. In the *Yale Law Review,* Tom Tyler offers an in-depth description as follows:

> The "CSI effect" is a term that legal authorities and the mass media have coined to describe a supposed influence that watching the television show CSI: Crime Scene Investigation has on juror behavior. Some have claimed that jurors who see the high-quality forensic evidence presented on CSI raise their standards in real trials, in which actual evidence is typically more flawed and uncertain. As a result, these CSI-affected jurors are alleged to acquit defendants more frequently...The perceived rise in acquittals can also plausibly be explained without any reference either to watching CSI or to viewing crime dramas more generally. For these reasons, and because no direct research supports the existence or delineates the nature of the CSI effect, calls for changes to the legal system are premature. More generally, the issues raised by current attention to the CSI effect illustrate the problems that arise when proposed changes in

> the legal system are supported by plausible, but empirically untested, "factual" assertions.[49]

In brief, the *CSI effects* have implications for how suspects are perceived in the legal system. Increasingly, the need for forensic evidence has become pervasive as it serves many purposes in jurisprudence and in the court of law globally. Fingerprints have helped in the quest for truth and justice. Genetic tests are equally construed to serve as biomolecular fingerprints in forensics. Genetic tests may serve as exculpatory evidence in the courts of law. It can also serve as inculpatory evidence in tying culprits to crimes and thus bringing about legal justice and this is even critical where the life of another person hinges on it. Since the late 1980s, molecular biologists and law enforcement experts have touted the idea of using DNA or genetic profiles to investigate and establish paternity disputes and criminal cases. In fact, basic electrophoresis DNA test results were acceptable as credible and reliable within the scientific and healthcare communities. DNA fingerprinting was used in resolving an immigration dispute in England. The immigrant was from Ghana and the UK immigration official initially raised question about the identity of the boy. The mother had three kids in Ghana but was not very certain about the actual paternal identity of one of the boys in question before immigration official. Therefore, Sir Alec Jeffrey, a molecular biologist was consulted to scientifically verify the biological identity of the boy in question. Taking blood samples from the mother and the three kids, he mapped out the DNA profile of the four minus the missing dad. Through a careful study of the test results, Jeffrey identified the similarity in the DNA profile of the three kids and their mother and the similarity of three kids linked to their dad. Evidently, the three kids were genetically identical to one father then in Ghana. This is how Jeffrey described it:

> Now it was a really tough case, right, first we didn't have the father for testing, so we had the mother and the boy in dispute. Secondly, we didn't have the sisters for testing, over in Ghana. Third, the mother wasn't terribly sure who the true father of this boy was anyway, apparently there are two fathers, neither of whom were available. So this looked like mission impossible for us. However, what we did have were three undisputed children in the family. So what we could do was to take the mother, and these were all done on blood DNA, so we had the mother's blood, three undisputed children. And you could take the mum's DNA fingerprint and the DNA fingerprint of the three undisputed kids and reconstruct the DNA fingerprint of the missing father, by identifying paternal characters in the children not present in mum. So we now had DNA fingerprint of the missing father, DNA fingerprint of the mother and then the DNA fingerprint of the boy in dispute, and every single genetic character in that boy matched either mum or a character in the missing father.[50]

This incident became the first time a DNA or genetic test was deemed as a scientific evidence, in adjudicating a criminal and capital case in the court of law. In another incidence in 1986, Professor Alec Jeffrey again assisted a court in the United Kingdom by using DNA profiling in the investigation of a rape and murder case initially tied to a teenager. Though the suspect, Richard Buckland, initially confessed, the DNA profiling test result paradoxically became exculpatory evidence in exonerating him. The suspect Colin Pitchfork initially avoided the *en masse* DNA fingerprinting tests but when Jeffrey eventually conducted the tests, it was a match culminating in his conviction including his own later confession of the heinous crime. The Pitchfork case became the second incidence of the admissibility of DNA as evidence in the court of law as well as the first criminal case adjudicated with DNA fingerprinting. For the first time in the United States, the Circuit Court in Orange County, Florida, accepted DNA fingerprinting evidence to convict Tommy Lee Andrews of rape. In a case in forensic anthropology, Professor Jeffrey successfully identified the remains of notorious Nazi physician, Dr. Josef Mengele. Jeffrey extracted DNA from skeletal remains, obtained DNA from Mengele's widow and son, and genetically identified and confirmed the remains to be those of Mengele. Genetic test results are now routinely acceptable as evidence in the court of law even though there have been some legal challenges regarding the testing procedures and the integrity of the labs. The proliferation of the tests also brings into question whether a court of law could subpoena a DTC test result directly from the providers as evidence. Most customers who submit to DTC genetic tests have reasons other than forensics. If a DTC genetic test result is used without consent from the customer as evidence in the court of law, it could contradict initial intents of customers and undermines the quest for individualized or personalized approach to health care and privacy.

In addition, DTC genetic tests if made publicly available (in court documents), could also subjugate families to undue suspicions and public ridicule as it is very easy to apply bioinformatics tools to extrapolate their genetic risks and disposition toward some diseases. It is likely that most DTC genetic consent is restricted to the customer and the testing agency. Such a fiduciary relationship is deemed violated, if the testing agency submits DTC results devoid of the customer's consent. It could erode trust and inhibit the roles of the genetic tests in public health and clinical research. As DTC genetic tests are currently not standardized, a false or an erroneous result could unjustifiably implicate an individual to crime and undue persecution, thus serving as Achilles' heels in jurisprudence! In a startling report entitled *Strengthening Forensic Science in the United States: A Path Forward,* the National Research Council raised and discussed some of the challenges associated with the application of DNA forensic as evidence in the court of law. The council noted with some concerns that, "the forensic science disciplines currently are an assortment of methods and practices used in both the public and private arenas. Forensic science facilities exhibit wide variability in capacity, oversight, staffing, certification, and accreditation across federal and state jurisdictions."[51] In other words, there are discrepancies or lack of uniformity in which biospecimens are tested for forensic purposes even though there have been federal and state laws governing these facilities. Furthermore, forensic labs "Too often have inadequate educational programs, and they typically lack mandatory and enforceable standards, founded on rigorous research and testing, certification requirements, and accreditation programs. Additionally, forensic science and forensic pathology research, education, and training lack strong ties to our research universities and national science assets." In the case of DTC genetic testing entities, they do not disclose most of their operations including methods, certifications, and personnel, thus creating a quagmire of suspicions. The council's recommendations are worth considering in view of the proliferation of DTC genetic testing at this time. Two of these recommendations are as follows:

> Recommendation 2:
>
> The National Institute of Forensic Science (NIFS), after reviewing established standards such as ISO 17025, and in consultation with its advisory board, should establish standard terminology to be used in reporting on and testifying about the results of forensic science investigations. Similarly, it should establish model laboratory reports for different forensic science disciplines and specify the minimum information that should be included. As part of the accreditation and certification processes, laboratories and forensic scientists should be required to utilize model laboratory reports when summarizing the results of their analyses.[52]
>
> Recommendation 3:
>
> Research is needed to address issues of accuracy, reliability, and validity in the forensic science disciplines. The National Institute of Forensic Science (NIFS) should competitively fund peer-reviewed research in the following areas: (a) Studies establishing the scientific bases demonstrating the validity of forensic methods.[53]

Preliminary conclusion

If the objective toward personalized medicine and health care is to be attained, lots of collaboration between individuals with genetic tests results (from DTC) and from their physicians (to the extent permitted by HIPAA and other health-related privacy laws) will be needed. There is chorus of call for Federal protection and strengthening of GINA in order to ensure and encourage individuals' greater participation in the new thrusts toward the entire gamut of personalized medicine and health care. Finally, as Heraclitus once indicated, *all things are in flux; the flux is subject to a unifying measure or rational principle.* Science by its very nature is fluid and changes rapidly as discussed in the first chapter of this book. Society is adapting to these changes in genomics by offering an array of DTC genetic tests, sometimes to the chagrin of clinicians. As newer, more efficient, and perhaps "precise" technologies emerge, it may even be possible for individuals to *perform* genetic tests similar to the HIV and pregnancy test kits personally. The current practice of DTC genetic tests in Heraclitean terms is in a flux: federal and international as well as professional policies are already emerging to regulate it. Reasonable regulations will ultimately change the directions and features of the DTC tests. But it is equally important to dare and continue the practice of DTC tests with an openness to change. As in the words of Kierkegaard, *to dare is to lose one's footing momentarily. Not to dare is to lose oneself.* Daring to undertake a DTC genetic test certainly launches a patient to a state of flux in which the customer momentarily plunges himself to an oasis of clinical uncertainty; dire anticipation of tests results, whether the test will be positive for an abhorrent biomarker or not, whether the test is even clinically reliable and potentially actionable. Such uncertainty will continue to be sustained in as far as the DTC genetic tests corporations *satisfy* patients with excellent customer services and products. DTC genetic tests transfigure (even momentarily) the "patient" into a "consumer" in providing diagnostics services hitherto limited to specialized clinics. This approach veils the customer from the stress of genuine genetic counseling, physician approval, and an array of clinical and bureaucratic bottlenecks of diagnostic services at the hospital. Regardless of the accuracy or scientific validity, DTC genetic tests have in no small measure muscled current informatics and other technologies that are easy, commercially accessible, and individually orientated. One of the pathways toward personalized medicine seems to be cusped under the aegis of the DTC genetic tests and symptomatic of a new threshold in patient care as a customized service. In a word, *DTC* genetic test has a role albeit amorphous at this time in achieving these goals!

End notes

1. Hippocrates. *Affections. Diseases 1. Diseases 2.* Translated by Paul Potter. Loeb Classical Library 472 (Harvard University Press; Cambridge, MA, 1988).
2. Hippocrates.
3. S.P. Mattern. *Galen and the Rhetoric of Healing* (Johns Hopkins University Press; Baltimore, MD, 2008). See also, S.P. Mattern. *The Prince of Medicine: Galen in the Roman Empire* (Oxford University Press; London, 2013) and Jeanne Bendick. *Galen and the Gateway to Medicine* (Bethlehem Book; Bathgate, ND, 2002).
4. Galen of Pergamum. The best doctor is also a philosopher, *Galen: Selected Works.* Translated by Peter N. Singer (Oxford University Press; New York, 1997) cf. Nancy G. Siraisi. *Medieval and Early Renaissance Medicine: An Introduction to Knowledge and Practice* (University of Chicago Press; Chicago, IL, 1990).
5. Nancy G. Siraisi. *Medieval and Early Renaissance Medicine: An Introduction to Knowledge and Practice* (University of Chicago Press; Chicago, IL, 1990).
6. Louis Pasteur. *Extension of the Germ Theory to the Etiology of Common Diseases* (Generic NFL Freebook Publisher; New York, NY, 1822–1895).

7. Ibid.
8. https://ghr.nlm.nih.gov/primer/testing/directtoconsumer
9. www.ghr.nlm.nih.gov
10. http://www.health.harvard.edu/newsletter_article/direct-to-consumer-genetic-testing-kits
11. C.S. Bloss, N.J. Schork, E.J. Topol. Effect of direct-to-consumer genome wide profiling to assess disease risk, *New England Journal of Medicine* 364(6): 2011, 524–534.
12. D.F. Easton. How many more breast cancer predisposition genes are there? *Breast Cancer Research* 1: 1999, 14–17; P.M. Campeau et al. Hereditary breast cancer: New genetic developments, new therapeutic avenues, *Human Genetics* 124: 2008, 31–42; and K.V. Voelkerding et al. Next-generation sequencing: From basic research to diagnostics, *Clinical Chemistry* 55: 2009, 641–658.
13. D. Carlisle. Testing times for BRCA gene carriers, *Nursing Standard (2014+)* 30(11): 2015, 17. See also, B. Chojnacki, R.F. White. The BRCA gene patents: Arguments over patentability and social utility, *World Medical & Health Policy* 5: 2013, 276–300.
14. Martina C. Cornel, Carla G. van El, Pascal Borry. The challenge of implementing genetic tests with clinical utility while avoiding unsound applications, *Journal of Community Genetics* 5(1): 2014, 7–12. See also, P. Borry, M.C. Cornel, H.C. Howard. Where are you going, where have you been: A recent history of the direct-to-consumer genetic testing market, *Journal of Community Genetics* 1(3): 2010, 101–106.
15. R. Dorfman. Falling prices and unfair competition in consumer genomics, *Nature Biotechnology* 31: 2013, 785–786. See P. Borry et al. Legislation on direct-to-consumer genetic testing in seven European countries, *European Journal of Human Genetics* 20(7): 2012, 715–721.
16. Ibid.
17. Y. Erlich et al. Redefining genomic privacy: Trust and empowerment, *PLoS Biology* 12(11): 2014, e1001983.
18. A. Macieira-Coelho. *Historical and Current Concepts of the Mechanisms of Aging* (Springer; Berlin, 2003).
19. Ibid.
20. Mudivanselage et al. Cost of living with Parkinson's disease over 12 months in Australia: A prospective cohort study, *Parkinson Disease* 2017: 2017, 13pp.
21. S.L. Kowal et al. The current and projected economic burden of Parkinson's disease in the United States, *Moving Disorder* 28(3): 2017, 311–318.
22. F. Jonasson et al. Macular corneal dystrophy in Iceland, *Eye* 3: 1989, 446–454.
23. Ibid.
24. Ibid.
25. The key to curing disease could lie in Iceland's genes—CNN.com
26. Editorial. Burst bubbles. Two medical-technology companies illustrate the ups and downs of innovation, *Nature* 526: 2015, 609–610.

27. Amy B. Vashlishan et al. Illusions of scientific legitimacy: Misrepresented science in the direct-to-consumer genetic-testing marketplace, *Trends in Genetics* 26: 2010, 459–461.
28. Amy B. Vashlishan et al. Illusions of scientific legitimacy: Misrepresented science in the direct-to-consumer genetic-testing marketplace, *Trends in Genetics* 26(11): 2010, 459–461; P. Kraft, D.J. Hunter. Genetic risk prediction—Are we there yet? *New England Journal of Medicine*, 360: 2009, 1701–1703; N. Wade. Genes show limited value in predicting diseases, *New York Times* April 15: 2009; M. Corpas et al. Crowd sourced direct-to-consumer genomic analysis of a family quartet, *BMC Genomics* 16: 2015, 910 and C.R. Lachance et al. Informational content, literacy demands, and usability of websites offering health-related genetic tests directly to consumers, *Genetics in Medicine* 12: 2010, 304–312.
29. Ibid.
30. Ibid.
31. How Reliable are Home DNA Ancestry Tests? Investigation Uses ....
32. Malorye Allison. Regulation of consumer genomic tests remains in limbo, *Nature Biotechnology* 27: 2009, 10. See also M. Foster et al. Evaluating the utility of personal genomic information, *Genetics in Medicine* 11: 2009, 570–574.
33. H. Howard, P. Borry. To ban or not to ban? Clinical geneticists' views on the regulation of direct-to-consumer genetic testing, *EMBO Reports* 13: 2012, 791–794; K. Kaphingst et al. Patients' understanding of and responses to multiplex genetic susceptibility test results, *Genetics in Medicine* 14: 2012, 681–687.
34. Alta Charo. On the road (to a cure?)—Stem-cell tourism and lessons for gene editing, *New England Journal of Medicine* 374: 2016, 901–990. See also, M. Araki, T. Ishii. Providing appropriate risk information on genome editing for patients, *Trends Biotechnology* 34: 2016, 86–90.
35. Ibid. Charo, 2016.
36. Hyoung Won Choi. Hereditary neuropathy with liability to pressure palsies, *Pediatric Neurology Briefs* 29(11): 2015, 83.
37. www.eeoc.gov
38. Sally Lehrman. Medical tests cost Lawrence Berkeley $2.2 million, *Nature* 405: 2000, 110. See also *Norman-Bloodsaw V. Lawrence Berkeley Laboratory* (*Case* No. 96-16526).
39. Sally Lehrman, 2000.
40. M. Otlowski et al. Genetic discrimination: Too few data, *European Journal of Human Genetics* 11: 2003, 1–2.
41. Francine Kauffmann. Tracing biological collections between books and clinical trials, *JAMA* 299: 2008, 19.
42. Jo Marchant. Ancient DNA: Curse of the Pharaoh's DNA, *Nature* 472: 2011, 404–406.
43. Alice Parker. Ten ideas that are changing the world, *Time Magazine* March 12: 2009.
44. Rogelio Lasso. *JAMA* 304(8): 2010, 909.
45. Ibid.

46. See also, N.A. Holtzman. Clinical utility of pharmacogenetics and pharmacogenomics In M.A. Rothstein (Ed.), *Pharmacogenomics: Social, Ethical, and Clinical Dimensions*. Hoboken, NJ: John Wiley & Sons, 2003, 163–186.
47. Dr. Randall Burt in Alice Park. Biobanks, *Time* March 12, 2009.
48. P.R. Billings et al. Discrimination as a consequence of genetic testing, *American Journal of Human Genetics* 50(3): 1992, 476–482.
49. Tom Tyler. CSI and the threshold of guilt: Managing truth and justice in reality and in fiction, *Yale Law Journal* 115: 2006, 1050–1085.
50. www.dnalc.org/view/15107-The-Ghana-Immigration-case-Alec-Jeffreys
51. National Research Council. *Strengthening Forensic Science in the United States: A Path forward* (The National Academies Press; Washington, DC, 2009).
52. Ibid. p. 22.
53. Ibid. pp. 23–24.

SECTION III

Regulatory Policy, Law and Biotechnology

4 The Global Regulatory Pathways of Biologics

An introductory comment

Biotechnology has contributed significantly to a robust biopharmaceutical industry and health care globally. As noted earlier on the biotechnology of history, biologics such as antibiotics have been part of health care over a millennial, and recombinant DNA (rDNA)-based biopharmaceuticals are increasingly popular in our time. Biologics such as serum and vaccines (either attenuated or nonattenuated) have been part of the backbone of the global biopharmaceutical industry and in the US in particular. Vaccines, a form of biologics, have been clinically used to ameliorate highly infectious and communicable diseases such as diphtheria, cholera, tetanus, tuberculosis, and a host of others for over at least a century. Indeed, biologics and public health campaigns have made it possible to curtail many childhood diseases and thus decreasing infant mortality and increasing longevity. The WHO, UNICEF, US Centers for Disease Control (CDC), and many other reputable health and allied health-based organizations have promoted mass vaccination programs for eradicating many diseases with biologics. The recent HIV and Ebola outbreak and many others continue to unravel the existential challenges to develop the next efficient generation of biopharmaceuticals in combating these. However, the current generations of biotechnological biologics are unique and different. These new generation of biologics are recombinant DNA-based and typically considered complex to produce. Nonetheless, they are significant and in high demand in treating many chronic and debilitating diseases such as cancer and Crohn's disease, to mention a few. Like chemically synthesized pharmaceuticals, biologics are subject to regulatory and policy approvals prior to marketing to the public.

The current generation of biologics as noted above, are recombinant biotechnologically based. They are often made through a medley of complex biotechnological process. Typically, the *process* involves the use of cells or complex biologic system—proper identification of genes in the host biologic system as well as the target disease is crucial. As such, a biologic or biopharmaceutical is generally considered an intellectual property. Given the fact that the biologics industry generates substantial economic value, investors raucously protect their patents and products. For the biotechnology sector, both the *processes* and the *products* are of great value. In a capitalist economy market driven by exceeding competition, profitability, and monopoly, it is incumbent on competitors to protect their business investments and ventures. These often cause some torsional strains and legal attritions of product infringements and perceived violations of rights among inventors and entrepreneurs of biologics. In particular, the new generation of recombinant biologics derived from human tissues and cells or biopsies, DNAs, RNAs, and others have led to so many patent infringement and data litigations and ethical discussions. Science as we have noted is a dynamic process. Therefore, biotechnological innovations especially recombinant biologics present very unique and rapid challenges for regulators unlike the chemical industry.[1]

Recombinant DNA products are expensive unlike new chemical entities (NCEs). This seems to be an obvious and forgone conclusion. The entire process of identifying the gene of interest, the host biologic system, the personnel (scientists and technicians), equipment, the uncertainty of the regulatory climate, and others translates into high costs associated with biologic medicine. In addition, the "generic" versions or "biosimilars" are relatively few as most referenced biologics have some data and market exclusivities, so the biologic market remains stagnant. Indeed, as Erwin Blackstone and Joseph Fuhr noted in *The Economics of Biosimilars*, "The high cost of pharmaceuticals, especially biologics, has become an important issue in the battle concerning ever-increasing healthcare costs. The average daily cost of a biologic in the United States is $45 compared with only $2 for chemical (small-molecule) drugs.[2] It is not surprising to hear the high cost of biologic drug for a year." For example, Marathon Pharmaceuticals recently approved biologic for treating *Duchenne muscular dystrophy* (DMD); Emflaza is pegged at $89,000 per year!

In addition, the conception that a "product" maybe extrapolated from a human part (such as the cell, DNAs, and tissues), and get patented and sold with a market and monetary value seem to define comprehension especially within a section of the public. Indeed, such a notion raises certain anthropocentric questions—What is the value of a human person? Can a human being be valued in monetary terms? Could the sale of a DNA or gene of interest extracted from a particular person be tantamount to quantifying and albeit undermine human value and dignity? As I will expatiate very soon, these questions emerged during the HeLa controversy. It seems many people want the HeLa cell lines, which continue to generate substantial financial value discontinued, as it appears to be violations of her rights and the dignity of her progenies since no expressed consents were ever sought from HeLa to monetize her biopsy. In other words, there seem to be a consensus to regulate biotechnology products due to the intrinsic nature of both processes involved in the manufacturing and the products especially those derived directly from human biologics.

The emergence and development of biologics regulatory pathways in the United States

Clinical research as a matter of categorical and regulatory imperative must align with international and local ethical considerations from preclinical,

clinical, and post-approval clinical testing and research due to some precedent about the mistreatment of research subjects. It is a reality that regardless of their sources or origins, recombinant molecular drug candidates are eventually tested on research subjects during the clinical phase of the research. Given some precedents, such as the Nuremberg Trials, the Jewish Chronic Disease Hospital scandal in Brooklyn, and the infamous Tuskegee Syphilis Project, a number of international and local ethical norms have emerged to guide the ethical conduct of researchers and sponsors during clinical research. As the General Principles of the Code of Helsinki (#10) notes: *Physicians must consider the ethical, legal and regulatory norms and standards for research involving human subjects in their own countries as well as applicable international norms and standards. No national or international ethical, legal or regulatory requirement should reduce or eliminate any of the protections for research subjects set forth in this Declaration.* As a result, clinical research must mitigate any harm and risks to research subjects. The DOH also recommends that proper consent from subjects must be sought prior to and during the research and confidentiality adhered to at all times. Furthermore, the clinical research protocols must be approved by an Ethics Committee (that is an IRB) in *...consideration with the laws and regulations of the country or countries in which the research is to be performed as well as applicable international norms and standards but these must not be allowed to reduce or eliminate any of the protections for research subjects set forth in this Declaration* (General Principles # 23). The *DOH* also calls for the protection of vulnerable research subjects.

The emergence of biologics regulatory pathways

Prior to 1901, there were no well-defined, standardized, or regulatory procedures for manufacturing biologics to ascertain their *purity* and *potency* as we have now. Nonetheless, these biologics were popular purportedly in ameliorating many medical conditions. However, two separate incidents in the United States changed the regulatory barometer culminating in the federal regulation of biologics and biopharmaceuticals. In 1901 in St. Louis, 13 children tragically died from a contaminated antitoxin serum made from a horse named Jim to treat diphtheria.[3] In another incident, nine children were reported to have died from another biologic: a contaminated smallpox vaccine in Camden, New Jersey. In response to these incidents and the need for further protection of public health and safety, the US Congress in 1902 promulgated the "*Biologics Act*" or the *Virus-Toxin Law* under the auspices of the Hygienic Laboratory of the Public Health and the Marine Hospital Service to address these challenges. The laws invested the regulatory authorities the legal power to license pharmaceutical companies and laboratories in order to manufacture biologics for human consumption. Public safety was obviously a heightened priority in these regulatory apparatus. Four years later, the US Congress also passed the Pure Food and Drugs Act. This legislative instrument prohibited mixing foods with drugs with preponderance false claims of medicinal value—a practice that was common at the time.[4] In 1938, the Federal Food, Drug and Cosmetic Act (FD&C)

was promulgated with specific provisions for the then National Institute of Health to regulate biologics. It was followed by the Public Health Service (PHS) Act (1972), which in pertinent parts empowered the FDA to regulate the production and licensing of biologics in the United States.

In brief, several circumstances and incidents (albeit tragic) involving biologics especially in the early part of the twentieth century led to the promulgation of many policies and laws to regulate biologics in order to protect public health and also to ensure the purity and safety of the products. These earlier forms of regulations have laid the foundation for a robust (bio) pharmaceutical industry creating the legal template for the involvement of society in the business of scientific research and innovation. Of particular interest in our discourse here is the intersection between biotechnology and regulatory policies. However, biologics regulatory norms remain relatively new and some in fact are still works in progress. This is partly due to the relative new technologies associated with recombinant biotechnology-based biologics. I think it is expedient at this time to offer some reflections on the nature and features of biologics and why these present opportunities for regulatory oversight in the public interest and, of course, in the advancement of science.

How did these regulatory laws define it? What are the constituent elements of a biologic? One of the earlier definitions encapsulated in Section 351 of the PHS Act, which in pertinent part states: a biologic is

> a virus, therapeutic serum, toxin, antitoxin, vaccine, blood, blood component or derivative, allergenic product, protein (except any chemically synthesized polypeptide) or analogous product, or arsphenamine or derivative of arsphenamine (or any other trivalent organic arsenic compound), applicable to the prevention, treatment, or cure of a disease or condition of human beings. [And it] can be composed of sugars, proteins, or nucleic acids or complex combinations of these substances, or may be living entities such as cells and tissues. Biologics are isolated from a variety of natural sources—human, animal, or microorganism—and may be produced by biotechnology recombinant methods and other cutting-edge technologies. Gene-based and cellular biologics, for example, often are at the forefront of biomedical research, and may be used to treat a variety of medical conditions for which no other treatments are available.

This definition by extension includes biosimilars or the "generic" versions of the referenced biopharmaceutical. More narrowly, biologics are large protein complexes typically cloned in cell culture through recombinant biotechnology means and latter purified for therapeutic purposes such as Humira (used in treating Crohn's disease). In brief, the regulation of biologics in the United States is under the auspices of the PHS Act. This includes the regulatory or legislative instruments covering the identification and characterization of the molecular candidature, research and development, manufacturing, data collection, approval, filling of the Biologic Licensing

Application (BLA), pharmacovigilance, and a gamut of other factors coterminous with the process of developing a biologic product that is safe, pure, and potent for human consumption or as clinically efficacious. These regulatory provisions ensure the purity and safety of biologics for the public and protection for the biotechnology companies. However, advancements of molecular biology especially in the area of molecular genetics and genomics are driving a new impetus for the next generation of biologics through rDNA biotechnology processes. The completion of the HGP has invariably infused a new perspective especially PM or precision medicine into the lexicon of biotechnology. And this new thrust undoubtedly is helping in a better understanding of the pathways undergirding many diseases as well as designing the new generation of biopharmaceuticals or pharmacogenomics that will be safe and potentially tailored to meet the individual needs of patients or a specific population target.

Indeed, the emergence, development, and the manufacturing of the new generation of biopharmaceuticals have significantly contributed to health care and the improvement of life of patients. In particular, biologics have seen substantial infusion of capital, R&D due to their intrinsic potencies to cure very rare and debilitating diseases, and as alternatives to chemically synthesized pharmaceuticals. R&D in rare diseases such as Crohn's seem to be receiving the needed attention and a renewed sense of urgency due to the advancement of biotechnological methods to find cures or develop therapeutic interventions. As Robert George once noted: "Much genetic knowledge has been generated by inquiry aimed at curing diseases, healing afflictions, and ameliorating suffering. Valuable biotechnologies have been developed for the purpose of advancing human health and well-being."[5] Biologics seem to hold the key in the global development of a globally robust pharmaceutical industry.

Features of biologics and biotechnology research

As indicated in the definitions above, biologics comprises nucleic acids, polypeptides in complex combinations through recombinant biotechnology means in living systems such as cells, tissues, or *in vivo*. In a word, biologics are large complex molecules such as proteins (polypeptides) in living systems. For example, genes from human biopsies may be genetically engineered to elicit immune responses to produce antibodies at specific loci in the human body and purified for therapeutic and research purposes. A typical biologic might have as much as 1,000–30,000 molecules biologically engineered to produce a biopharmaceutical in a living system, whereas a chemically based synthesized pharmaceutical such as Ibuprofen is composed of 33 atoms (C13H18O2 with a molar mass of just 206.3 g/mol approximately)! Because the molecules are typically large and complex, they posit many operational and procedural challenges to characterize and replicate even using the same biologic system from the same vial and sources. This is because living systems such as cells and tissues are dynamic with complex biologic networks: variables of vessels, intracellular matrix, protein–protein interactions, adaptability to internal and external ambient temperatures and response to other stimuli gives credence to the unpredictable nature in the manufacturing of biologics in general. In addition, several studies have shown that it is also difficult to replicate the same biologic products in the same living system under the same or similar conditions even in the same facility. Because of the complexity of proteins, modeling and designing poses substantial biotechnological challenges for bioengineers in the process of the production of biopharmaceuticals for human use in contrast to chemically synthesized pharmaceuticals.

Another issue is the structure, folding, and posttranslational modifications of proteins. Protein folding is determined by the amino acid sequences in living systems. Since there is a correlation between the structure of proteins and functions, proteins in biologics systems must fold properly in accordance with their 3D native structures in order to function biochemically or properly.[6] Granted that the modeling is successful, proteins must be correctly expressed in the biologic system, *translated,* and conform to the native proteins precluding denaturation or modifications. Any change or posttranslational modification could render the protein nonfunctional.[7] Hence, meticulous and consistent conditions of pH and temperature and other conditions in the production facilities are desirous for optimal production of biologics that are pure, identical, and therapeutically efficacious in every patient. This becomes even complex when a new cell line in a new vial has to be used to produce a biologic or in the process of making biosimilars (which will be addressed later in this chapter).

Furthermore, the processes in the manufacture of biologics are unique and of huge significance in biotechnology unlike chemically synthesized pharmaceuticals. As the aphorism goes regarding biologics—the process is the product![8] The complex nature of developing biologics is inherently linked to the entire manufacturing process. Therefore, the "process," which includes the identification of the molecular candidates through to the final product (upstream to downstream), culture, manufacturing, harvesting, purification, packaging, transport, and storage are of significant regulatory and commercial values. Every minute procedure in the manufacturing process from cell culture, harvesting, purification, characterization, and commercialization of biologics is carefully documented. These guarantee product consistency and purity at all times with almost negligible iota for error. Obviously, the biologic manufacturing process is an intellectual property and trade secret intrinsically, and biotechnology corporations and regulatory bodies have keen interest and fiduciary obligations to protect it.

Furthermore, a biologic product may be designed typically with a well-defined method or process. But another company or even the same company could develop a different or identical method with highly similar product for the same clinical indication. Nevertheless, maintaining such a consistency in terms of the purity and potency of the biologic is a huge issue in the field of biotechnology and scientific regulations. This is because biotechnology "methods" or "process" unlike mathematics or other inventions are considered intellectual properties. For example, Amgen received the patent covering the *exogenous* or artificial methods for the production of erythropoietin—a genetically bioengineered biopharmaceuticals through recombinant means in animal cells while Genetics Institute Inc., Cambridge, Massachusetts, patented the *endogenous method* in the production of Epogen (EPO) an essential biopharmaceutical in the treatment of anemia and also used in sports. In essence, both biotech companies were producing *identical* or highly similar and therapeutically efficacious biologics for the same clinical indication with *different methods*. This example also affirms the significance of the values of the processes of biotechnology research itself.

In addition, the entire process for the development of a biologic is highly regulated and egregiously expensive in terms of R&D compared to the production of chemically based synthesized pharmaceuticals. On an average, the cost of building biologic facility is estimated at $300 million, whereas chemically based drugs are pegged at $50 million. Comparatively, biologics are capital intensive even at the very early stages of development. This initial overhead costs alone could be a disincentive for many investors and startups. In addition, highly skilled scientists, technicians, and equipment (sometimes including robots) are required to manage these facilities and equipment throughout R&D, manufacturing, packaging, storing, and transportation even at the point of dispensing the biologics for human use. Such investments (both human capital, machineries) and the unalloyed regulatory requirements for consistency in biologics pose operational challenges for manufacturers.[9] Indeed, the cost differential in addition to some of the challenges above are typical features that biopharmaceutical companies have to navigate in order to meet regulatory, operational goals, and commercialization.

Besides, biologics are regulated both internationally and locally given some of the historic contexts discussed earlier in which some of the regulations were promulgated. Purity, potency, safety, and clinical integrity are key regulatory components that transcend the regulatory enterprise globally. There are also diverse international standards within the European Union (EU), United States, and Asia (though there is generally accepted move toward harmonizing these regulatory systems). Biologic development has become synonymous to a funnel. As several studies have shown, even though thousands of lead molecular candidates maybe identified during the initial stages of discovery, R&D, and in the pipeline, relatively few make it to the corridors of the approval Phase (III) in the form of a biopharmaceutical due to high attrition rates. It is therefore not surprising that experts and people question the rationale behind the attrition rates. Are these attrition rates justified? Why do regulators eventually approve relatively few and for the most part only one molecular candidate among the hundreds of thousands in the funnel of the drug development pipeline? How do companies write off their huge R&D financial expenditures estimated at over $1 billion per molecular candidate? Are these attrition rates ethically justified? To respond to these questions, it is important to examine a penumbra of factors and the very nature of the regulatory framework for biologics.

The prospects of biosimilars

Biologics or biopharmaceuticals are not new per se as indicated in my introductory comments on regulations of pharmaceuticals in the United States. To reiterate, following several incidents, the PHS Act of 1944 (PHS Act) were acted into law to regulate the production of biologics for human use. However, as with many regulations, contexts and circumstances change overtime due to many factors. Our healthcare system and (bio) pharmaceutical industries are dynamic and strive to incorporate the most updated and cutting-edge scientific innovations, technology, and skill to the profession. The new thrust in molecular biology and in particular recombinant biotechnologies continuously elucidate the frontiers for the development and the emergence of a new generation of biologics and of course these give vent to the FDA and the regulatory industry to create and monitor appropriate clinically approved pathways toward the development of a robust biologics industry. The molecular composition of biologics and biosimilars are therapeutically protein based and produced in living entities such as bacteria, viruses, cells, tissues, or *in vivo* unlike *NCEs* they have been regulated differently. However, biotechnology-based recombinant biopharmaceuticals generally are under the aegis of almost the same regulatory pathways and phases like *NCEs* counterparts but obviously with some ancillary changes. In addition, biosimilars are produced under the Biologic Act of 2010 (an amendment of the PHS Act of 1944), which was promulgated recently.

Clinical research on biologics are identical to general trends as their chemical entities, although there have been changes after the promulgation of the *Biologic Act* of 2010. So, how does the biotechnology process work? As an intellectually based industry, biotechnology relies on the best scientific ideas and turns these into products. Biotechnologist may think or conceive about an idea and design a theoretical framework for it. Next is to "prove the concept" or demonstrate that the scientific idea is plausible by conducting basic assays or experiments in the laboratory, individually or collaboratively. For example, the scientists may theoretically and experimentally demonstrate that several biologics such as cytokines or molecular entity has a therapeutic value. Typically, a scientist or cohort of scientists may hypothesize that a molecule or moiety has pharmacological value for humans. Therefore, after conducting several assays *in vitro* or/and *in vivo*, this molecular entity becomes a *leading candidate* to conduct clinical trials. After proper characterization of the molecule, a sponsor must demonstrate that in addition to the pharmacological activity in the molecule, it does not pose any acute toxicological danger to humans. Biotechnology production materials, which are living systems such as cells, cytokines, viruses, and tissues, are dynamic and subject to change and are capricious (unpredictable). Cell lines, tissues, or viruses used in running bioassays or even upstream production are subject to challenges such as contamination or impurities or many other biologic factors. Due diligence is required and if possible, genotyping of the cell lines and analytical characterization are incorporated into the production process unlike their chemical counterparts. As a result, maintaining the integrity and consistency of these systems are significant as the least change could alter the clinical or therapeutic value of the final product. In essence, the "process of biotechnology" itself is of huge economic and patent value to researchers and entrepreneurs.

Thus, following the successful identification of the molecular candidates, the FDA then requires the investigator/s to file an Investigational New Drug (IND) Application as the first stage required to start any test *in vitro* and *in vivo* or animal models in conformity with Section 21.C.F. R§ 312.23. Sponsors must demonstrate that the test is "reasonably safe" by including pharmacological data (including intended clinical indications) and toxicity testing. The IND dossier must contain proper identifications of the investigators/researchers including their qualifications and clinical expertise. In addition, the manufacturing process also exemplifies this notion of the process being the product. Therefore, after the product is well characterized, investigators collaborate to design the processes or methods to manufacture the product at the laboratory level and scale it up in order to yield products of high quality and quantity and of therapeutic value that is pure. Due diligence is paid to Good Laboratory and Clinical Practices even at this level of the development of the product. Pursuant to 21 C.F.R. § 312.23, the molecule must be determined to be reasonably safe. After this *...the sponsor then focuses on collecting the data and information necessary to establish that the product will not expose humans to unreasonable risks when used in limited, early-stage clinical studies* and establish the therapeutic and commercial value worth an R&D. In addition, because sponsors will likely ship the molecular compound *interstate* to other scientists and agencies for research purposes, federal law requires them to apply for an IND to the FDA for review and approval. The IND then marks the formal fiduciary relationship between sponsor and the FDA. Sponsors must demonstrate and fulfill three important data quadrants in the IND application about the molecular or drug candidate namely, animal pharmacology and toxicological studies, manufacturing information and clinical protocols as well as investigator information. All of these are important pharmacological information designed by the FDA to ensure that the biopharmaceutical candidate is safe, pure, and efficacious for human use during clinical studies. Sponsors must wait 30 days to hear from the FDA. The IND must have information on pharmacokinetics (PK), which entails absorption, distribution, metabolism, excretion, and half-life (ADME). In addition, pharmacodynamics (PD) demonstrate therapeutic value of the biologics especially its mechanism of action when used. Toxicological information is required—like NCE, a new biologics entity (NBE) IND must have substantial scientific data gleaned from assays on carcinogenicity, mutagenicity, and tetragenicity.

Consequently, if the FDA approves data from laboratory and animal studies encapsulated in the IND, the new biologic moves into Phase I of clinical trials and the molecular candidate officially tested in humans in controlled clinical studies. The main focus of Phase I is safety and dosage of the biologics in volunteers. On an average, 20–80 volunteers may be recruited into this phase according to current FDA regulations and current research practices. Healthy volunteers are normally recruited for this phase to demonstrate safety. Sometimes, patients with the target biologics are recruited. One of the prerequisites to start testing the biologic in humans is that the design must be approved by an IRB that it is compliant with the FDA and International Conference on Harmonizations (ICH) requirements and standards. Safety as the FDA noted: "…means the relative freedom from harmful effects, direct, or indirect, when a product is prudently administered, taking into consideration the character of the product in relation to the condition of the recipient at the time." Safety is a key component here so, information on the biologic activity and effects on research subjects are highly monitored. Toxicological and PD data from animal studies paves the way for the biopharmaceutical to enter *Phase I* clinical trial, which is an important milestone in the development of biologics as the drug candidate is tested in humans for the first time. Information on PK (effect of the body on the drug) is collected, documented, and analyzed. Specifically, it addresses questions on absorption, distribution, metabolism, excretion/elimination (ADME) of the biologic when tested on volunteers. Another data oscillates on PD that is the "effect of the drug on the body." Biologics drugs unlike chemically synthesized counterparts are proteins or DNA based so, their dosage and levels of absorption by the body especially intestinal poses operational challenges. Hence, the most preferred methods are intravenous or injection typically by a physician or a designated healthcare professional. Therefore, data on respiration and heart rate are recorded during the administration of the molecular candidate to determine the maximum tolerated dose (MTD). MTD generally focuses on what quantity the average human body can take. This includes the body's reaction to the pharmaceutical as the levels of concentration are increased until a safe MTD is attained. In addition, any *adverse events* (E) are determined at this phase. So, any unusual changes in the volunteers PD and PK could lead to abrupt stoppage of the clinical trials as the safety of the individual is key component in developing the biologic. According to several studies, approximately 10% of drug candidates in the preclinical phase makes it to this phase while 70% molecular candidates may pass this stage to the next phase. It should however be noted that attrition rates here are usually high. This poses some ethical challenges in terms of the costs of R&D and the determination of biologics pricing.

Furthermore, the main purpose or therapeutic objective for developing the biologics becomes the main focus of Phase II. Hence, the *NBE* is tested on homogeneous and small sample size population (20–100 subjects) for therapeutic or clinical efficacy of the target biologic. For instance, if the NBE's target is to cure breast cancer or pancreatic cancer, sponsors and investigators will recruit subjects with these clinical indications for Phase II. Data on clinical efficacy of the biologic for the target disease is determined and are carefully generated and documented. Furthermore, Phase II clinical studies also determine the common short-term effects of the biologic in actual patients. Since every drug potentially carries some modicum of risks (either known or unknown), a calculus of tolerable risk is determined during the control studies. Because of these factors enunciated above, Phase II clinical studies are highly monitored by investigators. It is estimated that 33% of biologics in the pipeline may move from this phase into Phase III.

The patient/research subject population in Phase III is diversified or heterogeneous and increased to between 1000 and 3000 to obtain sufficient information about the biologic risk–benefits relationship. The patient population for the study is randomized, placebo, controlled group, double-blinded and may take place at various locations concurrently. This helps determine whether the *NBE* clinically performs better than established or standard therapeutic care. The controls are also necessary for timely approval of the biologic as a biopharmaceutical for the indication under consideration. Further information of the efficacy of the biologic are generated and compared with the cross section of the research subject to prevailing or standard clinical treatments. Accordingly, subjects are grouped into two: prevailing clinical treatment of the indications and the NBEs target indications. Data from both groups are comparatively analyzed to determine which of the products and procedures will be best for the target indication. Sometimes, a situation called equipoise (i.e., a state of genuine clinical uncertainty) may arise where investigators are clinically unable to determine which of the two biopharmaceuticals or procedures are superior. Under such situation, clinical trials for the target indication have to stop. If there is no equipoise, then the studies move to the next level of obtaining a BLA for approval and marketing of the product.

In addition, this information would be used to label the biologic if approved by the FDA and for post-approval pharmacovigilance. It marks the official end of clinical trials. Sponsors may then submit the Biologics License Application (BLA) dossier for the approval of their NBE based on favorable reviews by the FDA. The biopharmaceutical for approval must have substantial data to demonstrate that it is "pure, potent and safe" for human use or as indicated for the target indication in the clinical study and as indicated in the BLA. The BLA must have information on the biologics' physiochemistry, microbiology, pharmacology, statistical, biopharmaceutical or biologics, and clinical data. In addition, information on the manufacturing processes the facility, storage, and packaging in compliance with *good clinical practices* (GCP) and ICH E6 requirements. Once approved, formal marketing and data protection begins after the typical 60 days wait. If approved, then Phase IV clinical studies must commence to demonstrate long-term effects of the biologic. Often this takes place in the form of *observation* of the actual clinical use of the biopharmaceuticals on real patient in clinical settings. Data are also generated for post-approval monitoring or pharmacovigilance. Currently, 12 years of data protection plus additional exclusivities are granted for biologics approved in the United States. The above regulatory process is confined to only the innovated or reference biopharmaceuticals, which spans not less than 12 years to manufacture. For the most part, pharmaceuticals have been synthesized from small chemical entities or NCEs for therapeutic uses. However, the emergence and the development of recombinant biotechnologies continue to redefine the scope of the pharmaceutical industries.[10] Biotechnology-based drugs are large complex proteins made in cells or living organisms. But there is no doubt that pharmaceuticals constitute significant portions of the overall healthcare cost in the United States and globally. In view of the above factors, branded biologics are expensive and often out of reach for many patients or populations. Others include international and local patent protections, market and data exclusivities, and other regulatory and approval exclusivities. For instance, Avastin is estimated to cost $50,000 while Herceptin costs about $60,000 per year; and Cerezyme costs $200,000, respectively. As a precedent, generics have played an important role in driving down costs associated with branded pharmaceuticals. But why do we have a relatively small generic version of biologics (biosimilars)? Does the regulatory process play a factor? To respond to this, I will briefly reflect on the current regulatory trajectory of the biosimilar.

As a way of recapitulation, branded or referenced biopharmaceuticals and protein-based drugs are generally expensive to manufacture due to a number of factors enumerated and discussed above.[11] Currently, it is estimated that between 70% and 80% of prescription medications in the United States are generics.[12] According to a recent study, generics alone saved the healthcare system and consumers in the United States; about $1 trillion between 2002 and 2011.[13] With an estimated per capita income of $53,143, it could be inferred that many patients in the United States might not be able to afford these lifesaving biologics.[14] A recent report compiled by the IMS Health suggests that "... from 2013–2018 generic drugs are expected to account for 52% of global pharmaceutical spending growth, compared to 35% for branded drugs. Overall, sales of generic drugs are forecast to increase from $267 billion in 2013 to $442 billion in 2017, an annualized growth rate of 10.6%." The seismic growth of 10.6% is partly due to what has been an important biotechnology semantic and legal phenomenon known as "patent cliff." A patent cliff is expected to occur because many buckbuster referenced *biopharmaceuticals* are heading toward the end of their patent life. Consequently, there will be neither data nor regulatory exclusivities, which create a curvature for competitors and biosimilar drugs (of course, unless there is a new indication). Given the seemingly astronomical costs of referenced biopharmaceuticals, a *patent cliff* is important with implication for consumers if competitors flood the market with their alternative or biosimilar products. During the launch of Zarxio (a biosimilar to Neupogen), reputed to be the first in the United States, Dr. Ralph Boccia, the Chief Medical Officer for the International Oncology Network (ION) said, "While biologics have had a significant impact on how diseases are treated, their cost and co-pays are difficult for many patients and the healthcare budget in general. Biosimilars can help to fill an unmet need by providing expanded options, greater affordability and increased patient access to life-saving therapies." Biotech companies may try throughout the patent life of the referenced biologic to "replicate" it, but may have to wait until data expiration occurs. So, a patent cliff will create a huge gulf of opportunity for the biotech companies to compete for, because data on referenced biologics become a public domain information. It is like after the party; everyone is offered the "exclusive" cake without any restriction. Because data are typically hard to find on branded biologics, biotech companies compete fiercely and whoever is able to find and demonstrate the molecular formulae and the process for the referenced biologic may prevail in applying for a biosimilar license and data exclusivities. As a result of the possible proliferation of replications of the same biosimilar, the costs of the biologic typically drop.

To address these challenges, the U.S. Congress passed the Biologics Price Competition and Innovation Act (BPCIA) herein referred to as the "Biosimilar Act" as part of the "Affordable Care Act" (ACA), which President Obama signed into law.[15] The law, analogized on the Hatch-Waxman Act, is meant to create an *abbreviated pathway* for generic versions/copies of biologics or NBEs in order to eviscerate the high costs associated with protein-based drugs and the overall healthcare system.[16] But has the "Biosimilar Act" actually created an abbreviated pathway for the development of biosimilars? Could the 12-year market and data exclusivities act as catalysts in ebbing or driving up the cost of biologics in general? What are the preponderance ethical and policy implications? How does the law balance the cost of innovations of biologics with lowering the price on biosimilars? This section examines these challenges of the "Biosimilar Act."

Generally, bioequivalent for generic versions of NCE pharmaceuticals are easier to replicate and formulate. While the regulatory pathways for NCE generics are well established, NBEs seem to be a niche in the process. This is due to a number of factors—the biologic market especially the new generation of recombinant biopharmaceuticals are relatively new and have considerable patent life and exclusivities still exists. However, a number of biologics have just reached patent cliff and at the verge of losing other market and regulatory protections globally. While the EU has a well-established biosimilar regulatory framework and market, in the United States, the regulatory and legal framework is still embryonic giving a new impetus for the emergence of the biologics. One of the prerequisites to start testing the biologic in humans is that the design must be approved by an IRB that it is compliant with the FDA and International Council for Harmonization of Technical Requirements for Pharmaceuticals for Human Use (ICH) requirements and standards. As indicated earlier, Congress just passed the Biologic Act as part of Obamacare, clearly defining the framework and investing the legal authority with the FDA for the regulatory process for biosimilars.

What then is a biosimilar in the context of the law? What level of similarity is required? What is this process? What are the requirements for getting a biosimilar approved? Does the policy require a clinical trial? What kinds of data are required and why? As indicated above, NCEs are regulated by the PHS Act of 1944 (PHS Act, unlike *new biologic entities*, which are regulated by the Food, Drug and the Cosmetic Act. While Hatch-Waxman (1984) regulates generics, the Biologic Act (2010) amended the PHS Act to create the pathway (expected to be abbreviated) for the biosimilar market. In order for a biologic to be approved in the United States, innovators are required to file form 351(k) to merit regulatory standards. The biosimilar or the follow-up biologic must be a "single biological product licensed under section 351(a) against which a biological product is evaluated." The biologic copy equivalent is often described as a biosimilar, follow-up biologic, biogenerics among others. Biosimilars are required to be chemically and therapeutically identical to the referenced biologic. Indeed, under Section 351(i)(2), *"biosimilar" or "biosimilarity" means that the biological product is highly similar to the reference product notwithstanding minor differences in clinically inactive components, and there are no clinically meaningful differences between the biological product and the reference product in terms of safety, purity and potency of the product* and...*the applicant (or other appropriate person) consents to the inspection of the facility that is the subject of the application.* This broad definition implies a high standard almost impossible to attain literally and cursorily. But a meticulous analysis of the syntax of the definition emphasizes that the biosimilar ought to be "highly similar" but not the same to the referenced product as with generics.

In addition, several key features could be extrapolated from the above and the policy in general. Biosimilars unlike generics are not generally identical to their respective referenced products. The biologic must be "highly similar" to the *reference product* (herein the original biologic) with ancillary differences in the clinically inactive component(s) of the molecule. This is to ensure that there are no clinically meaningful differences between the biosimilar and the referenced product or the innovator drugs—this is because any dissimilarity could alter the structure and potentially the functions of the biosimilar. Thus, the biosimilar must exhibit the same mechanism of action and the same route in terms of administration, dosage, and therapeutic potency. For example, Zarxio, the first biosimilar to the referenced Neupogen has the same route of administration. Therefore, the biosimilar can only be used under the same condition of the referenced or innovator biopharmaceutical or as indicated in the original BLA dossier and labels. Under current statutory regulatory provisions, in order for a product to be considered a biosimilar in the United States, innovators must *demonstrate* biosimilarity by providing data from laboratory studies, preclinical and clinical data. According to the FDA, sponsors must provide *analytical studies* data demonstrating that the biological product is *"highly similar"* to the reference product notwithstanding minor differences in clinically inactive components or moieties. Sponsors must have data to demonstrate proper analytical characterization of the molecular constituent and moieties of the biosimilar product, so this information is crucial. As there is a correlation between structure and function of biologics, hence the laboratory studies must demonstrate similarity of the structure of the biologic to the reference product to ensure that any minor differences do not impact the therapeutically relevant or active component of the molecule or moiety. Sponsors must include in the analytical studies, data on the expression system used, the manufacturing process, evaluation of physiochemical and immunochemical properties, functional activities, impurities, stability of the biosimilar in terms of structural integrity among others.[17]

Preclinical animal study data such as toxicological studies and PK are also required. These data are critical for the initiation of clinical studies because it provides information about the safety of the biosimilar in view of the reference product. A clinical study or studies (including the assessment of immunogenicity and PK or PD sufficient to demonstrate safety, purity, and potency in one or more appropriate conditions of use for which the reference product is licensed and for which licensure is sought for the biosimilar product. This is usually incumbent on the differences of the analytical, functional, and animal models between the biosimilar and the reference product. Typically, PK- and PD-generated data comparatively analyzed to establish if any meaningful clinical differences exist or/support the extent of "similarity" between the biosimilar and reference products. Documentation of product comparability is the last stage in conformity with 21 CFR 601. 12 or 21 CFR 314.70(g). Biosimilars are submitted under 314(k) of the PHS Act as amended by the ACA abbreviated pathway.

However, biosimilars, though theoretically plausible, are pragmatically nonexistent due to the unique and dynamic nature of biologics. Establishing similarities for biologics are generally akin to a herculean task because biologics as large protein molecules are difficult to replicate.[18] This is because biologics are made in living cells or in whole organisms and later purified, establishing "biosimilarity" could be inhibited by many factors such as differentials in temperature in bioreactors, protein folding or conformational changes, posttranslational modifications of proteins and the proclivities of genetic modifications of host organisms in altering the final products.[19] Indeed, studies have shown conclusively that "…the manufacturing processes cannot be duplicated."[20] Hence, a change of manufacturing facilities could potentially alter the biologic and biosimilars.[21] For example, the biologic, Epoetin manufactured in four countries were analyzed by researchers and they found some differences *in vivo* bioactivity ranging between 71% and 230% and substantial differences in their isoforms as well.[22] From these preliminary and yet important studies, it could be contested that unlike NCEs' generics, NBEs' biosilimars are not identical to their brand or original versions per se. Hence, the notion of establishing actual "similarity" and purity would be impossible to attain if consumer safety is to be guarded. Critics are quick to point out that since "true" copies of branded biologics are not possible, how could an "identical" biogeneric drug demonstrate the same potency? Could biosimilars in fact or in essence be interchangeable?

In addition, *interchangeability* of pharmaceuticals has been one of the catalysts for success of the NCE generic market globally. Pharmacists may interchange or switch biosimilar with referenced if the physician does not request so. These have many benefits and implications for health care. Healthcare analysts believe that just like NCEs, NBEs interchangeability could lead to a reduction in the cost of biopharmaceuticals to 20%–50% less of referenced drugs for patient care and insurance expenditures. Furthermore, essential biopharmaceuticals may become accessible to patients as the market expands with many competitor biosimilars that could be interchangeable. Pharmacists do have greater latitude, the regulatory, and the legal discretion to actually give the patient either an innovator pharmaceutical or the generic versions of a biologic. So, we often see the "*Do not Substitute*" box on prescription forms typically written by physicians. If a physician makes such a choice, he/she restricts the pharmacists from making any substitute for a similar or biosimilar (bio) pharmaceuticals for any medical indication. Questions have been touted and some consistent concerns expressed about the substitution of biologics in the wake of an emerging biosimilar market. Currently, for biologic to attain the status of interchangeability, it must be deemed and approved to be biosimilar to the referenced or innovator biopharmaceutical by the FDA. The FDA has issued additional guidelines and requirements for the consideration of a biosimilar approval as an interchangeable. According to the FDA, "An interchangeable biological product, in addition to meeting the biosimilarity standard, is expected to produce the same clinical result as the reference product in any given patient, and for a product that is given to a patient more than once, the risk in terms of safety and effectiveness of alternating or switching between the interchangeable and the reference product is not greater than the risk of using the reference product without alternating or switching."[23] In effect, innovators provide further data on PK, PD, toxicological studies, as well as data and evidence from some bioassays. The FDA guidelines recommend that innovators demonstrate through *bioassays* data the potency of the biosimilar product under consideration for interchangeability. The "process" for manufacturing unlike chemical entities is significant as it can have some effect on product quality. For example, a slight change in the *process* including source of materials and manufacturing plant could have some impact on the clinical potency and efficacy of the biosimilar.[24]

Another factor in establishing biosimilarity is biological integrity. Typically, biosimilars are not actual replicas of their referenced versions, so there is need to demonstrate or establish that candidates for molecular compound or moiety is of high resemblance and therefore biologically conforms to specifications in order to be deemed biosimilar. Innovators are required to conduct additional studies that include *in vitro* bioassays such as cell growth, enzymatic activity, antiviral, and infectivity assays; in addition, *in vivo* or assays in animal model and make this data available to the FDA. The amino acid sequence must be identical to the referenced biopharmaceutical. Any indication of heterogeneity or differences must be within acceptable regulatory thresholds. Data on hypersensitivity reactions (HSR) are equally important in order for it to be therapeutically alternative or interchangeable. In addition, PK studies in animals to determine the maximum concentration (C-max), the time taken by the biosimilar to reach maximum concentration (t-max), as well as the absorption rate of the biopharmaceutical and the time it takes to drop from its maximum concentrations. These data are critical in calibrating effective minimum concentration (MEC) in order to avoid drug resistance and increased risks of side effects. Furthermore, human pharmacological clinical studies data to demonstrate appreciable levels of immunogenicity, safety, and efficacy trials.[25]

Finally, this information must be compiled into a regulatory dossier "Documentation of Product Comparability" in conformity with 21 CFR 601.12 or 21 CFR 314.70(g) and submitted to the FDA for considerations. If all the above information is deemed acceptable, then a biosimilar could become interchangeable. And according to Section 351(k)(4) of the PHS Act as amended by the ACA, an interchangeable "biological product is a product that has been shown to be biosimilar to the reference product, and can be expected to produce the same clinical result as the reference product in any given patient." In addition, the biological product must be administered more than once to an individual and show that the risk in terms of safety or diminished efficacy of alternating or switching between use of the biological product and the reference product is not greater than the risk of using the reference product without such alternation or switch. If approved, a biosimilar becomes *interchangeable* to the innovator biopharmaceutical product and must be included in the *Purple Book* indicating whether "…a biological product licensed under section 351(k) of the PHS Act has been determined by FDA to be biosimilar to or interchangeable with a reference biological product (an already-licensed FDA biological product)."[26] These have many benefits and implications for health care. Pharmacists may interchange or switch a biosimilar with some referenced biopharmaceuticals even if the physician does not request so.

The regulator landscape in the United States is generally considered akin to the EU and other countries such as Canada and Japan. But each country seems to have some unique regulatory culture worth discussing in the light of biotechnological innovations. I turn now to reflect briefly on the emergence and the current biotechnological and regulatory landscape of Canada.

The biotechnology landscape in Canada[27]

The Canadian (bio) pharmaceutical industry bears enormous regulatory cloak as in the United States and the EU. There were similar historic precedents culminating in the direct involvement of the legislature (herein the Parliament) getting involved in regulating biologics a little over a century. Bio (pharmaceutical) production is a heavily regulated enterprise even though, historically, there were times that no regulation actually existed in Canada. However, certain historic incidents and precedents (tacitly indicated in the introductory paragraph to this chapter) have led to the need to regulate biologics, especially biopharmaceuticals for human use. The Canadian regulatory body analogized somewhat on the FDA and also emerged in "reaction" to certain unfortunate incidents involving science and tragedies in the latter part of the nineteenth and twentieth centuries in Canada.[28] The main instigative reasons and driving forces for the promulgation of the first extant regulatory norms (of significance to our discussion) were due to the adulteration of food and drugs for human consumption prior to 1909. Many people or "innovators" purportedly mixed chemicals with food and claimed these concoctions had medicinal and even magical values. In response, norms were initially enacted to merely regulate products on the market for human consumption at the time. However, the emergence of biologics such as serum, vaccines, and the continual adulteration of food as medicine led to the enunciation and the promulgation of the first extant regulator norm, the Proprietary or Patent Medicine Act (PMA) (1909) with explicit intent to regulate the production, licensing, and transport and sale of medicine within Canada.[29] The PMA requires the registration of all medicines produced in the provinces of Canada with obvious preponderance intent to protect public health, safety, and to ensure purity of products.

To bolster the regulatory oversight, an Act of Parliament, promulgated the Food and Drug Act (1920). The Food and Drug Act gave legal authority to the Health Ministry to officially *issue licenses* to manufacturers of pharmaceuticals. However, the authority and regulatory integrity of the laws came into focus during the thalidomide tragedy. This is because of the law's inability to prevent the tragedy. Accordingly, in 1951, the law required manufacturers to file New Drug Submissions (NDS) dossiers in order to receive the Notice of Compliance (NOC) from the regulatory body, Health Canada. They must also indicate evidence of efficacy in order to be deemed compliant prior to manufacturing and sales of pharmaceuticals, which also includes biologics.

Clinical trials are at the backbone of the bio (pharmaceutical) industry in Canada and obviously globally. Health Canada is the regulatory body responsible for the coordination of clinical trials. Health Canada (HA) has outlined the following reasons for regulating clinical trials[30]:

1. Protect the health of the people in the trial
2. Make sure the trials are well designed and conducted properly by trained professionals
3. Make sure that trials are monitored adequately and potential side effects are reported to Health Canada
4. Require that trials are reviewed by a Research Ethics Board

Akin to the FDA, Health Canada has a well-defined regulatory dossier to guide investigators and sponsors. It is worth noting that clinical trials in Canada, EU, and the United States are starkly similar though with minor procedural and terminological differences. For the most, they all have the preclinical, clinical (Phases I–IV) approval and postapproval monitoring. Of course data protection is also different in both countries due their respective intellectual property and juridical traditions. According to Health Canada, preclinical trials are considered under the aegis of the following:

I. The use of the drug in the patients being studied is appropriate
II. Any risk associated with use of the drug is minimized as much as possible
III. The best interests of the people participating in the trial are upheld
IV. The objectives of the trial are likely to be achieved

Phase I	(a) Determines safety (b) Safe range and dosage
Phase II	(a) Data on clinical effectiveness (b) Further assessment on safety (c) Determine the best dose
Phase III	(a) Monitor side effects (b) Clinical efficacy
Phase IV	Long-term benefits and postapproval surveillance studies

Final process approval: upon successful clinical trials, HA may weigh the *risks* and *benefits* of the drug in consideration. Typically, sponsors are anticipated to minimize the risks posts by a new pharmaceutical as much as possible. And the new pharmaceutical must be clinically efficacious. If the new drug meets these criteria, a sponsor will then apply for two documents: NOC and Drug Identification Number (DIN) in order for the drug to be approved in compliance with Part C, Division 8 of the Food and Drug Regulations. The issuance of NOC implies the new drug has met the safety, efficacy and quality of product standards of the Canadian regulatory norms. Sponsors will then apply for New Drug Submissions (NDS) dossier in order to sell or market the brand named drug to the public. The generics versions could be authorized for sale by the submission of Abbreviated New Drug Submissions (ANDS). Health Canada also permits the routes for marketing drugs with NOC. For instance, in case of mergers, change in renaming of a drug, a sponsor may so petition HC for authorization to market under the new name provided all the safety standards have been met.

The above regulatory framework, though generally applicable to chemically synthesized chemical pharmaceuticals, is by extension to the production of biopharmaceutical clinical trials as well. So, currently, the Biologics and Genetic Therapies Directorate (BGTD) is responsible for the regulation and the production of biologics in Canada. As one peruses the Canadian regulatory documents, one of the striking features is that there is no extant definition of "biologics." What one sees is a tacit *description* of what a biologic is and what it is not. The regulatory body indeed offers the following descriptions: *Biologics differ from other drugs for human use in that they must in addition to the information required for other drugs, include manufacturing information. This is necessary to help ensure the purity and quality of the product, for example, to help ensure that it is not contaminated by an undesired microorganism or by another biologic.*[31] As a result, "before a biologic can be considered for approval, sufficient scientific evidence must be collected to show that it is *safe, efficacious* and of *suitable quality.*"[32] The current biologics regulated by BGTD include "blood and blood products, hemostatic agents, bacterial and viral vaccines, hormones, enzymes, cytokines, monoclonal antibodies, allergenic extracts, gene and cell therapies, tissues, and organs."[33] These regulatory requirements are very significant and seemingly universal as they are found either directly or cryptically referenced in the FDA, EMA, and other regulatory instruments. Furthermore, biologics are to be placed on a "lot release schedule" according to *risks*. Indeed, question and index on risks of the biologic to human health is a serious issue in the Canadian regulatory system. As such, risk has been classified into three degrees:

1. *Higher risks*: Some biologics may pose serious risks to human health. As such biologics under this classification are tested before they are released for sale in Canada by the regulatory body in conformity to the norms.
2. *Moderate risks*: Biologics in this category are tested periodically or at the discretion of the regulatory body.
3. *Low risks*: Biologics in this category are not likely tested prior to being released for sale because they do not pose any known danger or there are no safety concerns due to available scientific data.[34]

In brief, the regulation of biologics and other pharmaceuticals is important in Canada. And Health Canada has designated authority and norms to ensure that safety, efficacy, and suitable quality of biologics are manufactured for the public. As such, manufacturers must as a categorical imperative adhere to these norms in order to access the Canadian market from the process of identifying a leading molecular agent, moiety through R&D to the pipeline, and obviously to the consumers. As an intellectual property, data protections including patent and other pertinent exclusivities are also regulated as they have significance for the biologics production landscape and overall healthcare needs of the public.

Data protection and patent are essential and well prized in biotechnology and the development of products, especially biologics. As a negative right, patents are easily quantifiable in monetary terms, hence each biopharmaceutical company protects it to the fullest extent of local and international norms. Currently, Canada grants 20 years of patent protection for both biologics and chemical entities. However, most of these years of the patent are curtailed due to lengthy and laborious processes involved in R&D of innovated (bio) pharmaceuticals. It is estimated that R&D of a single biopharmaceutical could take between 10 to 15 years (10–15) at about $1.3 billion.[35] This means that by the time R&D is complete and the biopharmaceutical is officially approved for human use, there is very little time left of the actual patent life. To compensate for this time lost and to incentivize innovations among others, the Canadian law and regulatory bodies follow the general global trend of protecting patents, data, and market exclusivities. The current practice is to grant eight years of patent protection and data exclusivities for both biologics and small chemical entities from the generics or biosimilars through its Food and Drug Regulations.

In addition, innovators could also apply for additional six months of patent protection or extension if there is a pediatric indication for the generic (biologics or chemical). This means that within this period, the innovator of the patent for the indicated (bio) pharmaceutical can challenge any direct patent or referenced patent infringement within the Canadian courts.

However, a comparative (bio) pharmaceutical submission is allowed within six years after the initial filling of data protection was granted to an innovator drug. In terms of "biosimilars" or subsequent entry biologics (SEB), unlike the United States and the EU, the Canadian law does not recognize this as innovations or novelties. Accordingly, neither patent nor data protection exclusivities are extended to cover biosimilars. Patent protection for biologics, in particular, has significant implications for consumers because no generic versions or biosimilars will be manufactured until after eight years meaning any blockbuster biotechnology-based pharmaceutical will be very costly until the patent expires. Unfortunately, illnesses and medical conditions do not usually wait for cheaper generics. I will expatiate on these issues and challenges further.

Patent, data, and market exclusivities of biologics

The EU member countries also have a long-standing regulatory, intellectual property, and legal tradition. Patent life in the EU is currently 20 years, which is consistent with global trends and traditions. The patent life clock begins to tick on the date of filling of applications to the European Patent Office. In addition to Patent protection, the EU also grants exclusivity for preclinical and clinical trial data to innovators or referenced drugs in addition to market exclusivity or protection. Prior to 2005, data exclusivity vacillated between 6 and 10 years (depending on the EU country). However, after the year 2005, eight years of data protection are granted concurrent with patent life of the referenced product two years of market exclusivity. A competitor may access and use both preclinical and clinical data for filling regulatory documents only after the eight to demonstrate bioequivalence but cannot market the generic or biosimilar products until the 10th year. However, the referenced (bio) pharmaceutical may receive an additional one year for a "new indication" of clinical significance over existing therapies and this will further delay the sale of the competitor one until the 11th year, hence the cognate 8 + 2 + 1 rules.[36]

Bioethics and some regulatory hurdles

The ACA explicitly states that innovators must demonstrate in their Biologic License Application (BLA) that the biosimilars are "safe, pure and have the potency" with new clinical data unlike NCEs. This is to insure and ensure that the biosimilar products are indeed safe and thus protects consumers from any adverse effects. Also, like their NCEs counterparts, NBEs biosimilars must be "interchangeable" to the referenced (bio) pharmaceutical. To attain that status, the biosimilar must demonstrate the same clinical indication in any patient in terms of purity, safety, and efficacy, as the referenced biologic. Therefore, *NBE* competitors must conduct at least one clinical trial and collect data to demonstrate about "immunogenicity and pharmacokinetics or pharmacodynamics that are sufficient to demonstrate safety, purity, and potency in one or more appropriate conditions of use for which the reference product is licensed and intended to be used and for which licensure is sought for the biological product…."[37] Clinical trials, in general, can (and often) account for some of the huge costs (about $1.2 billion) associated with research and development (R&D).[38] Indeed, R&D for biologics is exceedingly higher than NCEs.[39] And because costs associated with R&D are indices and predictive of the market prices of drugs in general, biosimilars are not likely to be less costly in a short period of time. Therefore, while the ACA has created the pathways for biosimilars with anticipation of reducing healthcare costs in the long term from the policy perspective; after all, prescription drugs constitute between 10% and 15% of overall healthcare costs in the United States, asking for additional clinical data could potentially increase the costs of biologics and bioequivalents.[40]

The (bio) pharmaceutical approval process is laborious and requires copious infusion of capital to the extent that on an average it takes nearly 14 years for an FDA approval. This means that by the time of approval by the FDA, patent protection on biologics will have been running out. Data and market exclusivities granted to innovators are significant determinants for drug companies to gain market monopolies within a legally defined time framework so they could recoup investments and make some profits on their products. Unlike NCEs that have five years, NBEs are granted 12 years of market exclusivities.[41] As Gabrowski noted, "Data exclusivity provisions are therefore designed to reduce uncertainty and provide some stability and predictability for developers and investors against costly litigation and early patent disruption. They also provide an important incentive for products that spend a long time in basic research or clinical development after their core patents are filed. Novel products with new modes of action in particular often have lengthy discovery and development periods. Data exclusivity also encourages innovators to continue research and development (R&D) for new indications. This post approval research is an important pathway in biologics for enhancing patient health and welfare."[42] This continues to generate debates within the legislature and many stakeholders in the (bio) pharmaceutical industry. The cause of disagreement is that such a long period of exclusivity and market monopoly could drive and sustain a higher price tag on biologics due to a concomitant absence of competitors to the branded biologics.

Generally, patents are *negative rights* that act as stop signs for competitors from making and distributing intellectual property without a legitimate license or some form of legal or quasi written agreements.[43] As a negative right, patent data have been at the epicenter of much litigation due to infringements either directly or indirectly or in any form. As a result, non-patent holders and competitors do not have the right to use data under patent to develop the same or similar product without explicit permission in the form of licensing agreements or as so granted by a competent court with proper jurisdiction. This is significant for the (bio) pharmaceutical industry. Clinical research and patent dossiers are among some of the most valuable and protected data in the field of recombinant biotechnology-based pharmaceuticals globally. There are two types of data exclusivities that are typically granted. These are patent and market exclusivities. Of course, a biologic manufacturer could avoid filling for patent and keep such data as trade secrets but risk infringement and unauthorized disclosures. However, as a general rule of thumb, innovators usually will file for patent protection. Current patent laws in the US protect both the process and methods of biotechnology drugs. As the FDA has noted, "Issuance of a biologics license is a determination that the product, the manufacturing process, and the manufacturing facilities meet applicable requirements to ensure the continued safety, purity and potency of the product." The patent including drug information such as the process, methods, formulation, and equivalents become public domain information in the *Purple Book under the ACA analogous to the NBE Orange Book. The Purple Book* "...includes the date a biological product was licensed under 351(a) of the PHS Act and whether FDA evaluated the biological product for reference product exclusivity under section 351(k)(7) of the PHS Act."

In addition, the FDA also grants ancillary clinical exclusivities such as pediatrics pursuant to Section 505(A) for the moiety or the biologics among others.[44] Others include indications for rare and orphan diseases. Furthermore, because most approved biologics may still be within patent protection, innovators could additionally explore what has become known as "ever greening" to further extend data and market exclusivities. For instance, 46% of 60 biologics approved between 1986 and 2006 have received additional approval for other indications; Avastin (originally approved for metastatic breast cancer), was also approved for lung cancer. Humira, a $10 billion blockbuster biologic's patent expires by 2016.[45] However, a new moiety could be added for a *new clinical indication* or it could be reformulated for a *new administering route* in order to qualify for additional market and patent exclusivities. In other words, after 12 years, one could anticipate continual exclusivity and market monopolies.[46] Each day of market or patent exclusivity is worth many profit margins for the (bio) pharmaceutical industry. Nevertheless, the brunt of such lengthy exclusivity is not likely going to encourage potential competitors to enter the biologics markets sooner for which the law was passed.[47] This policy issue was raised during the debate leading up to the promulgation of the ACA.[48] Also, while the average development rates for NCE are about 90 months, it takes 97.7 months for NBE to be developed.[49] However, the approval rates for NBE are 30.02% in comparison with 21.5% for NCE. Phase I in clinical trials of NBE are longer (19.5 months) than NCE (12.3 months) even though the attrition rate of biologics seems a bit lower than chemical entities.

This implies that costs of R&D coterminous with NBEs may be higher than NCEs.[50] If so, then 12 years of market exclusivity may be justified. However, such justifications may undermine the intent of the "Biologics Act," which was crafted on the Hatch-Waxman Act in view of ensuring healthy competitions to branded drugs and curtailing the costs of pharmaceuticals. If 70%–80% of all prescription drugs are generics, then such lengthy monopolies could astronomically increase the cost of health care and render essential biopharmaceuticals not accessible to those who might be in urgent need of them.[51] As already indicated, R&D takes up a significant cost of biologic and biosimilar development due to many regulatory and scientific reasons and factors. According to Gabrowski, R&D is typically over $1 billion with high attrition rate. Hence, it is a financial gamble to begin with. In addition, clinical trials exceed a decade of research. Moreover, for biosimilars, a competitor must conduct additional trial to generate clinical data to meet current regulatory demands under the ACA. Neither an innovator's nor competitor's sponsor has automatic access to the market under the completion and evaluation of the new data that the biosimilar has the same clinical indication as the referenced in patients without any additional reformulation or changes. Thus, the new clinical trial will increase the cost for R&D unless in NCE generics. Analysts speculate that this could potentially stall the price of biosimilars rather than act as a catalyst for price reduction. Advocates are touting the idea of analogizing biologics and biosimilars to clinical trials for "orphan" and "rare" disease or as in the EU where no further clinical trials are required. In fact, the EMA has approved at least 20 biosimilars since 2006. This has created a robust biosimilar market and pipeline of access to patients and options to physicians and pharmacists. In a scathing report released by IMS Health recently, it was noted that biosimilar medicines have the potential to enter markets by 2020 for a number of key biologics that have current sales of more than EUR40 billion. Cumulative potential savings to health systems in the European Union (EU) and the U.S., because of the use of biosimilars, could exceed EUR50 billion in aggregate over the next five years and reach as much as EUR100 billion. Stakeholder choices expand with the availability of biosimilars, including increased patient access to the same molecule or other medicines: use of biologic treatments has increased by as much as 100% following the availability of biosimilars in the EU.[52] In view of this, there is a growing call for harmonization between the EMA and the FDA in developing a better and robust biogeneric industry that will meet global healthcare demands and needs: creating greater patient access to essential biopharmaceuticals at relatively affordable prices that generates enough revenue and profit for manufacturers.

Nomenclature, surprisingly, has become a bone of contention in biotechnology products. Generally, small molecules and large biological molecules (bio) pharmaceuticals are assigned nomenclatures called *International Nonproprietary Names* (INN) by the WHO.[53] The INNs are also their generic names. According to the WHO, the INN nomenclature "…is important for the clear identification, safe prescription and dispensing of medicines to patients, and for communication and exchange of information among health professionals and scientists worldwide."[54] In terms of biosimilars, the WHO recommended that both glycosylated and nonglycosylated be named differently. Glycosylated biosimilars such as Epogen have a Greek alphabet as a suffix to the INN of the reference drug while nonglycosylated biosimilars have the same name as the reference biologics. One of the major challenges facing the naming of glycosylated biosimilars is that the Greek alphabet may soon extirpate as more biologics get approved and that means a different nomenclature or taxonomy ought to be used, which could create some confusion among dispensers or pharmacists. And a wrong dispensing and use of a biologic could be horrendous for the patient. Nonetheless, the nomenclature becomes even critical during recalls of (bio) pharmaceuticals due to safety concerns and for the purpose of pharmacovigilance. Giving the significance attached to the INN nomenclature, the omission of following the INN nomenclature of biosimilars within the "Biologics Act" has been hotly debated. As noted above, the INN nomenclature has served a global purpose to the extent that both the FDA and the EMA have been using it for decades. The omission has cascaded in a labyrinth of ethical speculations and debates; this is so because it may compromise public health, especially if there are adverse effects coterminous with a biosimilar that is being recalled thus potentially subjugating patients to harm! As the President of the Generic Pharmaceutical Association (GPhA) succinctly noted: "...the INN approach to biosimilar naming has proven safe and effective in Europe, it has worked in the United States for chemical drugs and currently approved biologics, therefore it should be the standard for US biosimilars."[55] A change in the nomenclature could potentially compromise the safety of patients and undermine the global efforts of pharmacovigilance. Another implication is that it may be difficult to track down fake biosimilars especially in developing countries that relies heavily on the INN system for the healthcare needs of their population in stark contradiction of the ethical principle that enjoins on all and sundry not to cause any harm. After all, the FDA has a *prima facie* obligation in ensuring that all (bio) pharmaceuticals they approve are safe to the extent that all information about biosimilars are disclosed to the public.

As noted in the introductory paragraph, the innovation of protein-based biopharmaceuticals continues to change the landscape of the pharmaceutical industry and health care in general. The need for innovated drugs for the treatment of diseases is crucial in the overall health care and wellbeing of the citizenry. Some studies have suggested by 2016 that biologics will constitute about 17% of overall global expenditure on prescription drugs.[56] Ensuring that the health of the people is protected constitutes a moral imperative to craft laws that both incentivizes and rewards innovators of biologics for the public good. But does the "Biologics Act" actually protect the health of the people since most biologics are egregiously expensive far beyond the affordable capacities of many people? Was the concern for the health of the people the guiding principle leading up to the passage of the law or the concerns of the biopharmaceutical companies? In evaluating the debate leading up to the passage of the law, there is no doubt that the biopharmaceutical industry assiduously lobbied for the most part for the longer exclusivities and patent protections among others. While in reality, most biopharmaceutical companies by default have specific disease targets and pathways in developing biologics, most patients or the general public has been barely active in debating the issues leading up to the development and passage of the Biologics Act. In brief, this chapter contends that the health of the people should be the guiding principle in implementing these laws. If so, then patients or the public as a categorical imperative actively and fully involved in this debate to the extent that the ACA should be subject to some amendments that truly reflect their needs for quality but affordable health care. After all, as the title of the Biologics Competitive Act (BCA) suggests, it is an affordable healthcare act! The essence and the application of the policy should translate into the care of patients.

Health care especially in the United States has been tied with economic opportunities.[57] An individual's economic venture or employment defines the kind of health insurance coverage he/she may have access to. A good employment has been coterminous with good insurance coverage and health care. The euphemisms of the introduction of the ACA was to curtail the high costs associated with health care in the United States and also to guarantee that as many people as possible have health insurance and ensure that prescription drugs especially biologics become affordable. However, a synoptic purview of the NCE and NBE policies seem to suggest that the price monopoly on biologics is not likely to change until after 12 years. Even then, competitors have the prodigious task of demonstrating that their biosimilars are "equivalent" and "interchangeable" to the brand NBEs. It means that patients who might not have adequate insurance coverage literally have to be on the waiting bandwagon until after 12 years in order to afford the generic versions or biosimilars when prices would have automatically waned down assuming there are competitors. Could this cause some *harm* in contradistinction to the ethical dictum of *primum non nocere* and also lead to rationing of health care? Some states are currently debating as to whether "minorities" should have access to essential biologics such as Sovaldi because it is very expensive. As demonstrated above, the most vulnerable in society may suffer the brunt of such policies in stark contravention of the ethical norm that "above all, do no harm." Unlike the Hatch-Waxman Act, it seems the "Biologics Act" does expose the most vulnerable in society to the dictates and the dynamics of the market in terms of price tags on biologics. It has been contended that the high price on biologics could have synergistic effect on the most vulnerable and economically disadvantaged in society. This can affect economic productivity, thus posing additional burden on scarce national resources and individual patient who might need limited but egregiously expensive biopharmaceuticals to improve or sustain the quality of their lives.

The next ethical concern is about the rights of every human being to have access to adequate health care. The Universal Declaration of Human Rights Article 25 imposes a moral obligation on society to respect the rights and health of individuals. The declaration also accords each person the right to adequate health care. As a signatory to this charter, the United States has the obligation to protect and accord the rights of each citizen's access to adequate health care, which include lifesaving biopharmaceuticals. For instance, the cost per pill of the Hepatitis C biologic Sovaldi is $1000 per day with complete treatment cost estimated to be $86,000 with successful treatment rate of 90%.[58] Recently, Illinois enunciated some criteria for approving Hepatitis C patients in their Medicaid program for Sovaldi.[59] Only the most sick were qualified to have access to Sovaldi. More so, minorities were excluded under the expediency that "...these people were not included in [the] research that was conducted. We did not feel the drug [Sovaldi] was made for everybody with a diagnosis of hepatitis C…. So we could not see everybody getting a prescription, just because their own data says it's effective about 90% of the time."[60] While the intent of the "Biologics Act" was to address this phenomenon, such *medical rationing* of lifesaving biologics is jeopardizing the health of individuals who have a legitimate moral claim of rights to access health care. Furthermore, if each member of society has a right to health care, then society has the moral and fiduciary obligations in ensuring its' members have access to quality affordable health care including safe pharmaceuticals. This socioethical obligation may be inferred from *Jacobson v. Massachusetts* where the judges indicated: "the… [Constitution] laid down as a fundamental … *social compact* that the *whole people covenants* with each citizen, and each citizen with the whole people, that all shall be governed by certain laws for the 'common good,' and that government is instituted 'for the protection, safety, prosperity and happiness of the people, and not for the profit, honor or private interests of any one man.'" While the ACA was crafted to address these obligations, it is doubtful whether biosimilars would ever be affordable given some of the major hurdles discussed above. Should society begin to reconsider some other ways of supporting biologics development such as waiving fees associated with clinical trials and also grant five years of market exclusivities in order to curb the high costs so as to make them accessible and affordable?[61]

It is important to note here that innovators also have rights to protect their intellectual property and recoup the high cost associated with R&D of biologics. Balancing market monopolies and safeguarding patents especially in a globalized world poses some challenges. While there are legal avenues to litigate infringements especially in the United States, Canada, United Kingdom, and Japan among others, challenging infringements in other places could be significantly costly. Currently, pharmaceuticals in India have made biosimilar of Sovaldi at $2500 in contrast to $84,000 required for the entire treatment in the United States. This piece contends that if the price tag on biologics such as Sovaldi has been lower than $84,000, it was unlikely that such infringement could have occurred which could affect the profit margins of Gilead. Surprisingly, Gilead has agreed to lower the price of Sovaldi in India and other countries to as low as $9000 for the entire treatment of the disease. The ethical question herein is, if that is possible in India and Egypt why not in the United States? After all, the United States bears the burden of creating this novel drugs and yet its' citizenry cannot afford it! Is this ethically justified?

In addition, the World Trade Organization's (WTO) agreements on Trade and Intellectual Property (TRIP) also explicitly allows for infringement in the wake of "health emergencies" of epidemic proportions. WTO TRIPS (Article 31) states in pertinent part *…in situations of national emergency or other circumstances of extreme urgency, the right holder [of a patent] shall, nevertheless, be notified as soon as reasonably practicable*. The presumption that the high price of branded biologics coupled with the fact that the

"Biologics Act" has granted a lengthy data protection could lead to a tsunami of infringements of patents and market exclusivities across the globe is highly contentious. Such infringements may be justified under the ethical principle of utilitarianism. Utilitarianism suggests that in a moral conundrum, one must choose or act in a way that brings about the highest good or pleasure for the highest number of people. Patent infringement could bring about the availability of lifesaving and expensive biologics to patients evidenced with the Indian and the Egyptian productions of the (bio) generics to Sovaldi. The consequences are also manifold; biosimilars may get to the market earlier than 12 years at relatively lower prices. Since biologics are extremely complex to duplicate, the potentials of cheap and impure biosimilars may flood the market with a preponderance import of endangering patients and public health. The other side is that the biopharmaceutical companies may lose significant profit margin if their patents are infringed, which does constitute a moral injustice to them as well. Leveraging the calculus for protecting innovators' rights and less costly biosimilars will continue to pose ethical conundrums.[62]

In brief, this section has surveyed and examined biologic production from the scientific and regulatory and ethical perspectives. Biotechnological innovations hold great prospects in an array of areas especially in medicine. As a scientific discipline, biotechnology is a dynamic process. Due to precedents, the regulation of biotechnological products such as biopharmaceuticals have become normative and legally binding with many implications for us as humans. As an intellectual property, the process, methods, and products of biotechnology has also created some legal reviews and it is expedient at this time to explore some of the issues from a legal perspective.

End notes

1. Mark P. Matthieu. *Biologics Development: A Regulatory Overview* (Parexel Intl Corp; Waltham, MA, 2004).
2. Erwin A. Blackstone and Joseph P. Fuhr, Jr., The economics of biosimilars, *American Health Drug Benefits* 6(8): 2013, 469–478.
3. Matthieu, M. *Biologics Development*.
4. G. Milligan and D.T. Barret (Editors). *Vaccinology: An Essential Guide* (Wiley-Blackwell; New York, 2015).
5. Robert George. Ethics, politics, and genetic knowledge, *Social Research* 173(3): 2006, 1029.
6. Lehninger. *Principles of Biochemistry* 287–288.
7. Ibid.
8. Ibid.
9. M. Matthieu, *Biologics Development*.
10. Saurabh R. Aggarwal. What's fueling the biotech engine—2012 to 2013, *Nature Biotechnology* 32: 2014, 32–39.
11. Affordable Care Act. See also Henry Grabowski. Follow-on biologics: Data exclusivity and the balance between innovation and competition,

Nature Reviews: Drug Discovery 7: 2008, 479–488; Joseph DiMasi and Henry Grabowski. The cost of pharmaceutical R&D: Is biotech different? *Managerial and Decision Economics* 28: 2007, 469–479; Henry Grabowski and Richard Wang. The quantity and quality of worldwide new drug introductions 1992–2003, *Health Affairs* 25(2): 2006, 452–460.

12. *IMS Institute for Healthcare Informatics* (www.imshealth.com/institut) The *Use of Medicines in the United States: Review of 2010.*
13. Ibid www.gphaonline.org.
14. Ibid. See also www.fda.gov (*Implementation of the Biologics Price Competition and ...*).
15. Ibid. Grabowski. *Follow-on Biologics*. See also Mari Serebrov. WHO: Biosimilars not the same, why should names be? *BioWorld Today* March 28: 2013.
16. Henry Grabowski. Follow-on biologics: Data exclusivity and the balance between innovation and competition, *Nature Reviews* May 28: 2008; Duff Wilson. More cost cuts sought from drug industry, *The New York Times* July 22: 2009.
17. Rick Ng Drugs. *From Discovery to Approval* (Wiley-Blackwell; Hoboken, NJ, 2015).
18. Huub Schellekens. How similar do 'biosimilars' need to be? *Nature Biotechnology* 22(11): November 2004, 1357–1359.
19. Ibid.
20. H. Mellstedt.
21. Rick Ng. *Drugs: From Discovery to Approval.*
22. Ibid.
23. www.fda.gov.
24. Daan J.A. Crommelin et al. *Pharmaceutical Biotechnology: Fundamentals Applications* (Springer Science; New York, 2013).
25. Ibid.
26. www.fda.gov.
27. Kshitij K. Singh. *Patentability of Biotechnology: A Comparative Study with Regard to the USA, European Union, Canada and India* (Springer; New Delhi, India, 2014) and www.hc-sc.gc.ca.
28. Ibid. www.hc-sc.gc.ca.
29. Ibid.
30. www.hc-sc.gc.ca.
31. Ibid.
32. Ibid.
33. www.hc-sc.gc.ca.
34. Ibid.
35. Gabrowski.
36. Daan J.A. Crommelin et al. *Pharmaceutical Biotechnology: Fundamentals Applications*.
37. TITLE VII—Improving Access to Innovative Medical Therapies: Subtitle A—Biologics Price Competition and Innovation SEC. 7001. I (aa) (*Affordable Care Act*). See also www.healthcare.gov and H. Mellstedt. The challenge of biosimilars.
38. www.acrohealth.org.

39. Hamilton Moses. Researchers, funding, and priorities: The Razor's edge, *JAMA* 302: 2009, 1001–1002.
40. See Gabrowski.
41. Ibid Angell. *The Truth about the Drug Companies: How They Deceive Us and What to Do about It* (Random House Digital, Inc.; New York, 2005); R.G. Hill et al. *Drug Discovery and Development: Technology in Transition* (Churchill Livingstone; London, 2012). 2nd edition; Chapters 1–4: 14–17; 19–22.
42. Gabrowski.
43. See also John R. Thomas. Follow-On Biologics: The Law and Intellectual Property Issues, Congressional Research Council Report, January 15, 2014.
44. Ibid.
45. Grabowski.
46. John R. Thomas. pp 6–9.
47. Congressional Research Council.
48. Ibid.
49. Gabrowski.
50. Ibid.
51. Norman Daniel. *Just Health.*
52. IMS Health. *Delivering on the Potential of Biosimilar Medicines the Role of Functioning Competitive Markets*, 2016.
53. Council Guidelines on the Use of INNs for Pharmaceutical Substances, 1997.
54. Ibid.
55. Mari Serebrov. WHO: Biosimilars not the same, why should names be? *Bio World Today* March 28: 2013.
56. Patricia V. Arnum. Tracking growth in biologics: The share of biologic-based drugs in the global pharmaceutical market is on the rise, *Pharma Technology* February: 2013.
57. Ibid.
58. Ed Silverman. How illinois allocates $84,000 drug for Hepatitis C, *World Street Journal* August 3: 2014; Ed Silverman. How much? Gilead will charge $900 for Sovaldiin India, *Wall Street Journal* August 7: 2014.
59. Ibid.
60. Ibid.
61. Congressional Research Council.
62. Congressional Research Council.

5
Biotechnology in the Court of Law

Association for molecular pathology versus myriad genetics

Generally, society has been repugnant at the idea of patenting human genes and synthesized recombinant nucleic acids and biospecimens in biomedical research. Indeed, the evidence that genes are embedded with information and are shared by other family members seems to galvanize societal rejection of patenting them. As Deegan pointed out, *a gene patent is an intellectual property, which gives the patent holder the right to exclude others from making, using, selling, or importing an invention for a period of time....*[1] Should genes therefore be patented? Whose intellectual property—the specimen donor's, families', physician's, or the researcher's? Most importantly, what are the implications of the law on patenting of genes or genomic materials to pharmacogenomic and PM? The debates seem to hinge on certain significant legal questions: are genes products of nature or are genes human inventions? An affirmation of the former means that within patent jurisprudence, cDNAs may not be patent eligible while the later implies the patentability of genes because they are human inventions in view of Section 101 of the Patent Act. The third question is on the extent to which a researcher could claim *property ownership* on another person's genomic materials and biodata. Since patent laws in the United States have evolved, it is important therefore to explore how these laws are applicable in current jurisprudence and their significance for the advancement of PM, most especially the post-Myriad implications.

First, let us examine the question of patentability. Prior to 1980, Title 35 of the United States codes governed every iota of patent litigations. Section § 101 describes patent eligible matter and precluded *living organisms*. However, the *Diamond v. Chakrabarty*, 447 U.S. 303 (1980) when the Supreme Court in 5-4 decision ruled *inter alia* that live human-made microorganisms are patentable subject matter under Section 35 § 101 changed the legal landscape to include living organisms and derivatives such as genomic information and proteins. This change in the law cleared the legal conduit for patenting many biological and genomic materials that have become the bedrock for the emergence of a robust biologics industry. One of the earliest significant litigations on the patentability of genetic materials occurred in Washington State (*Association for Molecular Pathology et al. v. Myriad Genetics*). Mutations in the BRCA gene and its variants are responsible for nearly 7% of all breast cancers.[2] Myriad developed an assay for identifying the mutations in these genes (BRCA 1 & 2) and successfully filed for patents. The lawsuits challenged the validity of Myriad's *claims* on isolated genes, diagnostic procedures, and methods for identifying gene expressions and specific *drug targets* for the BRCA 1 & 2. Initially, the Federal Court in New York ruled that synthesized DNA de novo may be patent eligible. The court, however, was inconclusive on whether isolated cDNA (which does not have introns) was different from naturally occurring ones. The case was appealed to the Supreme Court of the United States for further consideration. The Supreme Court ruled overwhelmingly in a 9-0 vote that merely isolating genetic materials does not make it patent eligible pursuant to the Patent Act, Section 101. The presiding judge, Justice Clarence Thomas averred *inter alia*:

> Myriad did not create anything,... to be sure, it found an important and useful gene, but separating that gene from its surrounding genetic material is not an act of invention. A naturally occurring DNA segment is a product of nature and not patent eligible merely because it has been isolated. It is undisputed that Myriad did not create or alter any of the genetic information encoded in the BRCA 1 and BRCA 2 genes.

The court also invalidated the *methods* for the isolation of genetic materials in view of Section 101 of the Patent Act of the United States which in pertinent part states: *Whoever invents or discovers any new and useful process, machine, manufacture, or composition of matter, or any new and useful improvement thereof, may obtain a patent therefore, subject to the conditions and requirements of this title*. While Myriad lost parts of its patent claims, it nonetheless won claims for patenting synthetic or cDNAs. The decision also implies that researchers would no longer pay royalties and licensing fees to Myriad Genetics. It is important to note that following the ruling, many genomic diagnostic companies and laboratories immediately reduced the cost for testing for mutations in the BRCA gene. In fact, Gen DX, Ambry Genetics, DNA Traits concurrently reduced their testing fees for the BRCA genes from an average of $3000 to $995.[3] This is significant for pharmacogenomics development because researchers no longer have to fear any legal barriers for using information extrapolated from the BRCA genes to develop their own biologics or chemical entities in search for targeted/personalized therapy. In addition, the ruling is also significant for early GWAS studies or "population-based" screening, diagnostic, and personalized therapy for carriers of the genes—specifically, women and men with these genes could be screened as soon as possible for early therapeutic intervention, if necessary, to save lives. Physicians could integrate testing for the BRCA genes for the patients. The ruling has also paved the way for biomedical scientists to use the myriads of genomic data available for collaborative research in pharmacogenomics and pharmacogenetics development and personalized therapy. And as the outgoing Director of the FDA, Dr. Margaret Hamburg observed, "There is no single discovery...to address our unique set of modern scientific regulatory challenges. But one thing is clear: if we are to solve the most pressing public health problems we face today, we need new approaches, new collaborations and new ways to take advantage of 21st century technologies. And we need them now."[4] Clarifying the potential legislative and legal hurdles for a robust genomic medicine is critical and I believe the ruling seems to have done that albeit the controversies it continues to generate.

Some scholars have suggested that the dismissal of gene patents could potentially truncate innovations in molecular bioengineering, biologics, and clinical research.[5] Indeed, the completion of the HGP also means lots more genomic data are available for GWAS and the ability to synthesize genomic material for biomedical research has also significantly improved. Since most researches are privately funded (75%), it means companies or researchers will have to invest their own resources in their genomic projects in anticipation of a reasonable patent and financial protection.[6] Consequently, the ruling implies virtually little patent protection for their genomic researches and intellectual work. This could potentially serve as a disincentive in advancing private investment in genomic and pharmacogenomics researches if the field of PM is to stand the test of time. Michael et al. noted, "Privatization [patents] must be more carefully deployed if it is to serve the public goals of biomedical research. Policy-makers should seek to ensure coherent boundaries of upstream patents and to minimize restrictive licensing practices that interfere with downstream product development. Otherwise, more upstream rights may lead paradoxically to fewer useful products for improving human health."[7] Furthermore, the ruling could also obliterate the sharing of critical genomic data because researchers will have to protect their genomic data as an intellectual property under the aegis of trade secrets. This is because in exchange for patent protection, researchers typically will have to make their data publicly available, but with the court ruling, genomic data will have to be protected from the public. This could affect GWAS and lead to duplicity in genomic researches and potentially cascade in high costs of pharmacogenomics research and PM. Indeed, the opinion of the Myriad ruling seems to leave a pendulum of other nagging questions. First, the court opined that the criterion for the patent ineligibility of naturally occurring DNAs and cloned sequences or cDNA is that of their respective informational contents rather than their chemical structures. The court seems to suggest that isolated DNA in *toto* contains coding information while cDNA are essentially chemical structures of the former. The court seems to redact myriads *claims* that DNA essentially contains information. The court opined *inter alia*:

> Nor are Myriad's claims saved by the fact that isolating DNA from the human genome severs chemical bonds and thereby creates a non-naturally occurring molecule. Myriad's claims are simply not expressed in terms of chemical composition, nor do they rely in any way on the chemical changes that result from the isolation of a particular section of DNA. Instead, the claims understandably focus on the genetic information encoded in the BRCA1 and BRCA2 genes. If the patents depended upon the creation of a unique molecule, then

> a would-be infringer could arguably avoid at least Myriad's patent claims on entire genes (such as claims 1 and 2 of the '282 patent) by isolating a DNA sequence that included both the BRCA1 or BRCA2 gene and one additional nucleotide pair. Such a molecule would not be chemically identical to the molecule "invented" by Myriad. But Myriad obviously would resist that outcome because its claim is concerned primarily with the information contained in the genetic sequence, not with the specific chemical composition of a particular molecule.

This opinion as noted above creates a seemingly nebulous situation on the nature of DNA. This is because it seems to suggest that the *genetic information* encoded in the BRCA 1 and BRAC 2 genes is what makes naturally occurring ones patent-ineligible under the Patent Act but not the chemical structure of the DNA sequence itself. If these claims are accurate (as the opinion of the court seems to suggest), then cDNA sequences could equally not be patent-eligible because it is possible to create synthetic or cloned polypeptides of different chemical compositions and yet code for similar or same informational content. For example, according to research conducted by Jonathan Greene and others, the *Bacillus subtifis* phage SP0l contains hydroxyl methyluracil (HMU) in its DNA in lieu of thymine (T) in its homolog.[8] Generally, transcription factor 1 (TF1) according to Green, "binds selectively to HMU containing DNA while the bacterial HU/DBPII proteins are thought to bind DNA nonspecifically." But through recombinant biotechnology methods, researchers could clone this gene sequence and replace the HMU with thymine.[9] In other words, while these homologs may be similar in terms of their informational sequence, nonetheless their chemical structures and functions differ. But clone genes will actually be the same contrary to the court's opinion that "Such a molecule would not be chemically identical to the 'invented'…."[10] In addition, some patents have been issued for naturally occurring primers, iRNA constructs and short nucleotide probes. For example, a patent issued after the Myriad ruling to Alnylam Pharmaceuticals on a double-stranded RNA, targeting the inhibition of the expression of the Serpinc1 to treat hemophilia is still valid among others.[11] Thus, narrowing the litigation to the *informational* content of naturally occurring DNAs posited some intriguing questions. In perspective, one could suggest that Myriad and any pharmaceutical company would be concerned about both the chemical compositions as well as the informational content in any gene of interest either naturally occurring or cloned. Thus, both naturally occurring polypeptides and complimentary DNAs may be patent-ineligible and this could potentially be challenged in the court of law in future litigations.

Another central nagging question from *Myriad v. AMP* case is whether the decision is applicable to humans or are nonhuman genes also encapsulated in the ruling? Obviously, BRCA 1 and 2 genes referenced in the claims of the patents that constitute the patent litigations were specifically predicated on human genes responsible for some breast cancers. But the bone of contention before the court as noted in my introductory comment in this section was: "Are human genes patentable?" In response, Justice Thomas noted: *"This case involves claims from three of them and requires us to resolve whether a naturally occurring segment of deoxyribonucleic acid (DNA) is patent eligible under 35 U. S. C. §101 by virtue of its isolation from the rest of the human genome."*[12] In his concluding statement, however, Justice Thomas did not categorically align the findings of the court specifically to human genes or genomes. He dexterously notes in his closing adjudicative statement:

> Nor do we consider the patentability of DNA in which the order of the naturally occurring nucleotides has been altered. Scientific alteration of the genetic code presents a different inquiry, and we express no opinion about the application of §101 to such endeavors. We merely hold that genes and the information they encode are not patent eligible under §101 simply because they have been isolated from the surrounding genetic material.[13]

These concluding statements of the court leave a gulf of questions opened. Does Myriad's decision apply to isolated human genes or nonhuman genes or both? Specifically, does the ruling imply that naturally occurring DNA in other eukaryotes? We could infer from the premise of the litigations that the court was addressing specific gene mutations in humans and not nonhuman genes. If so, then though the concluding statements do not *ipso facto* mention "human genes" *as such,* it is so implied. But such conclusions based on presumptive inference rather than definiteness could be a recipe for a *Lake Wobegon fallacy.* But in case law, the ruling becomes a precedent for adjudicating future litigations oscillating on DNA and genetic products. Furthermore, previously issued patents on naturally occurring DNA may be considered invalid pursuant to the Myriad ruling and this could potentially increase the number of *inter pates reviews* (IPR) under the aegis of the Leahy–Smith America Invents Act especially in the United States.[14] In fact, Gene DX has since the ruling filed 11 *IPR* against myriads and other diagnostic testing companies. As of February 17, 2015, however, Myriad has settled these IPR.[15] Furthermore, the Myriad decision does not address patents on purified recombinants genes or polypeptides occurring in nature, which are cloned and synthesized in artificial carriers such as yeast and other vectors that are generally considered a bastion for the development of biologics and other biopharmaceuticals. To avoid this pitfall and litigations, biotech companies could *design* products and moieties *around* DNA and obviously avoid using words or expressions such as "discovering DNA" in their claims even though that could increase operational or R&D costs.[16]

In addition, there were many other speculations about the potential impact it may have generally on the biopharmaceutical industry as well as the development of PM. These speculations could be synthesized and summed up in the following assertions: first, that the decision will cascade in a *chilling effect* on the biotechnology industry in general and in particular the pharmacogenomics development. Also, it may serve as a disincentive for innovators whose intellectual property could no longer be protected by the patent law on biologics. And that the ruling may discourage or stall investment or capital infusion into the biotechnology industry. Fourth, it has been asserted that innovators may intend to keep innovations such as diagnostics as trade secrets, which potentially drive up costs such as having exclusive rights to critical tests needed to clinically validate actionability. But have these assertions come true after the Myriad decision? Let us take a critical analysis of these.

I believe it is worth discussing these issues emanating from the Myriad decision as to whether the ruling will have any impact on scientific innovation. Specifically, what are the economic impacts of the ruling on the biopharmaceutical industry and how could this affect the development of precision medicine? Also could the ruling potentially serve as a disincentive to pharmacogenomics innovations? On a positive note, the ruling on *Myriad v AMP* has resulted in the reduction of testing fees associated with BRCA assays and other genetic diagnostic tools. Myriad immediately reduced its testing fees to under $1000 from $ 2500.[17] Other diagnostics companies also followed suit in reducing their testing fees. License fees as well as royalties hitherto paid to Myriad for the BRCA diagnostic kits were obliterated. And this impact in particular seems to be the very driving force behind many advocates and policymakers and some researchers who have consistently postulated that the test should be accessible, available, and above all affordable especially when the probability of positive polymorph of the BRAC 1 and 2 genes are symptomatic of a high propensity for cancer. Furthermore, Myriad's stock appreciated moments by 13% when the ruling took place but percolated to 6% by the close of the trading day.[18] Could these market indices be merely an incipient reaction to the ruling just like many landmark biotech decisions? Indeed, other biotech companies with vested interests in pharmacogenomics and diagnostics such as Merck, Glaxo-Smith, and Amgen have also seen a surge in their shares and stocks. According to a research report, the biotech industry shattered records for venture investment, IPOs, and M&A in 2014 as growing enthusiasm for breakthrough technologies and rising stock prices drove investment. Overall, the global life sciences industry raised a total of $104.2 billion, up from $92.9 billion in 2013[19] and in fact … drug approvals, biotech stocks significantly outperformed the major indices, and the industry set records in many key financing categories.[20] This is contrary to the projection that the ruling may negatively impact the biopharmaceutical industry; it seems this has not been the case at least at the moment. The "post-Myriad impact" seems to be fudged since it cannot be substantiated by any convincingly statistical evidence of financial significance. As a matter of fact, diagnostic testing has surged especially for the BRCA genes and ancillary tests. These assessments of the impact of the ruling are only two years and perhaps time will tell. But going by Moor's law used as a predictive indicator for innovative development, perhaps the effects are negligible on the development of a robust pharmacogenomics industry. As Cook-Hegaan noted, based on extensive study, "Gene patents have proven useful in developing some therapeutic proteins; neither the harms nor the benefits of DNA patents for clinical genetic testing is clear" and finally fears that gene patents might impede scientific research have not been borne out, at least to date.[21] In brief, the stock market or the financial portfolios of most biotech companies have not been affected contrary to prediction. In response to the question of the ruling serving as disincentive to innovators; it will rather bolster innovators confidence and investments in isolated DNA products. As indicated, the biotech stocks are still doing well. Many laboratories are now offering the BRCA 1 and 2 tests at reasonable fees and this is obviously good for the development of precision medicine.

What about the chilling impact of the ruling on innovations? It was widely speculated that the decision may have a chilling effect and stall innovations in genomic science.[22] Generally, patents offer substantial incentives and protection for patentees. This assertion is medicolegally true to some extent because research and innovations with a patent has been the bastion of the biopharmaceutical industry. In essence, Myriad and other innovators have invested substantial manpower and investments in discovering and making the BRCA kits as well as marketing these products and services. In contradistinction to this view, an internationally acclaimed geneticist and a Nobel Prize winner, Dr. John Sulston suggests that patenting genes have "chilling impact on research, obstruct the development of new genetic tests, and interfere with medical care…rather than fostering innovation."[23] These assertions have also been bolstered by Dr. Francis Collins in these words: "The right to control exclusively the use of patient's genes could have made it more difficult to access new tests and treatments that rely on novel technologies that can quickly determine the sequence of any of the estimated 20,000 genes in the human genome."[24] While these statements are coming from very credible authorities, some others have also argued that innovators might resort to keeping their innovations under the aegis of trade secrets (unless reversed engineered by others). It is also important to point out that the Myriad decision did not obliterate all gene-based patents. In fact, the ruling was legally "narrowed" and the opinion, a terseness of only 18 pages continues to stir a turbid of discussion. In view of these conflating opinions above on the "chilling effect" post-Myriad, it is worth noting that more innovations in the biopharmaceutical industry have seen a surge with concomitant investments portfolio as even noted above. I believe these challenges will continue to emerge as experts continue to decipher the functions of genes of the entire genome in view of PM. In brief, the post-Myriad decisions have been and will continue to be fraught with a myriad of uncertainties as to the short or long term impacts on the development of a robust pharmacogenomics industry and precision medicine.

The next discussion is on the *ownership* of bodily parts and genomic materials/information. The question is who *owns* DNA or genetic information? Do individuals own their bodies as properties? If so, do they exclusively own their genomic information as well? What happens to discarded biopsy materials and inherent genomic information? Do they belong to the original owners even if they sign out consent forms relinquishing ownership? What about their immediate family members who share similar genetic materials? Should they be required to sign off consent forms if these materials are to be used for biomedical research? In addition, it seems clinical research on small compounds are drying up while research on biologics are seeing significant surge globally.[25] Which means more biological and specifically human specimens and genomic materials will be used in the process of drug development (pharmacogenomics). Undoubtedly, the age-old question of who owns an individual's body will continue to elicit concatenations of debates evidenced in *Moore v. Regent of University of California, Los Angeles.* John Moore had a *hairy cell leukemia-variant* (HCL-V) and went to the UCAL Medical Center for treatment. The attending physician, Dr. David W. Golde performed a spleenectomy and extrapolated other bodily parts and fluids for testing and analysis. Over a span of seven years, Moore went back to see Golde where various forms of tests and bodily fluids were taken but, unknown to him and without his consent, Golde had used his biological specimens to develop a cell line. Golde actually filed for a patent and had substantial financial and stock benefits from it. Moore filed a lawsuit averring that Golde did not seek for his consent and made claims for property rights and other compensational damages. The court ruled against Moore sending shock waves to advocacy groups who champion absolute consent for all biological specimens by researchers. This paper contends that in an era of PM, researchers must be transparent about their intents and possible use of genomic information they obtained from their patients/research subjects.

As noted above, the law on patenting genetic and genomic material is an evolving one. I wish to turn at this point to examine some legislative and policies on the extent to the status of the genomic information and symptomatic import for PM. Are there legislations at the federal and state levels that address these issues? What are the possible implications of these for PM? Several states over the past few years have recognized the debate and have introduced several legislative instruments to address the growing societal concern about the status of genetic and genomic materials. Other issues addressed include whether human biospecimens, and genomic data be considered property or vice versa. South Dakota introduced a bill, SD H.B. 1260, in an attempt to address some of the issues raised above. The legislative instrument in pertinent part states: *All DNA, genetic information, or results of any genetic test, as related to health, benefit plans, are the sole property of the person from whom it was derived. DNA, genetic information, or test results may only be used or acquired with the permission of the person tested or, if the person is under age eighteen, with the permission of a parent or legal* guardian (Section 2 paragraphs 9–12). It is worth mentioning here that the law categorically states genetic and genomic information "are the *sole property* of the person *from whom it was derived,*" which means no other person could lay claim to it or even file a patent for such information. In Alabama, the legislature made a similar assertion in their bill AL H. 78 stating that "genetic information is the sole property of whom it was taken." In Massachusetts, MA S.B. 1080 and Vermont VT H. 368 the legislatures also stated that genetic information is the *exclusive property of the individual from whom the information is obtained* (MA S.B. 1080, Lines 5–6). The legislative instruments in all of these clearly preclude another person other than the person from whom the genetic information is derived as the sole owner. Sole ownership (of genetic information) in common law is synonymous to making a property claim of an entity. Also, property is a quantifiable entity with commercial value. But can a human being be quantified as a commercial entity? I think the property arguments remain fluid because genetic materials transcend individuals. In other words, genetic materials are shared by people within families. For example, when scientists wanted to authenticate HeLa cells, they extrapolated biospecimens (with consent) from direct members of the Lacks family because they share the same genetic information. Genetic information is therefore not the "sole property" of the person from whom it was taken per se because all progenies share these as well. Furthermore, treating a human being and human specimens as property evokes an effervescence of slavery where other human beings purportedly owned others as their sole properties. These could potentially truncate GWASs since some people may not want to make their genetic information available for valuable scientific study and thus obliterate the efforts toward PM.

With the completion of sequencing of the entire human genome, it is anticipated that claims on genetic information as property will be unprecedented. But as the data on the human genome seem to suggest, there are so many genetic similarities that people within some population targets share with each other. A *claim* by an individual on his/her genes or genomic information *ipso facto* will be a claim on another person's due to their common genetic interrelatedness. These will invariably continue to create legal strains among competing claimants and it will be a categorical imperative on both society and the courts to find some modicum of pragmatic and operative agreements on construing human genes as properties. Some scholars have argued for absolute privacy in shielding individual genomic materials from the public. Others advocate for absolute openness and sharing of genomic information and thus allow research to patent them as well in an effort to bolster pharmacogenomics and PM.

In addition, as discussed in the first section of this chapter, greater confidentiality and protection of genomic data are critical in precision medicine. As the director of the NIH, Dr. Collins rhetorically posited "Ask people, 'Are you comfortable having this specimen used for future genomic research for a broad range of biomedical applications?'—if they say no, no means no."[26] This is a profound observation because it respects the autonomy of the individual to make his/her own decisions about genomic data as well as ensuring that such data could potentially be useful for society in general if allowed. But current genomic data does have implications for future progenies as seen in the HeLa cell line controversy. I believe these questions will continue to posit challenges for the court and obviously a mire of ethical and policy discussions as the new generation of recombinant tools such as gene editing continue to emerge toward the development of precision medicines and the development of a robust biologics industry.

End notes

1. Robert Cook-Deegan. Gene patents, *From Birth to Death and Bench to Clinic: The Hastings Center Bioethics Briefing Book for Journalists, Policymakers, and Campaigns* (The Hastings Center; Garrison, NY, 2008): pp 69–72.
2. Lasker Award Population-Based Screening for *BRCA1* and *BRCA2* in *JAMA* (September 17, 2014); see also Collin B. Beg et al. Variation of breast cancer risk among *BRCA1/2* carriers, *JAMA* January 18: 2008.
3. Jeffrey M. Perkel. Gene patents decision: Everybody wins, *The Scientists* June 18: 2013.
4. Margaret Hamburg. *Advancing Regulatory Science for Public Health* (2010): p 3.
5. Michael A. Heller et al. Can patents deter innovation? The anticommons in biomedical research, *Science* May 1: 1998.
6. Justin Chakma, Gordon H. Sun et al. Asia's ascent—Global trends in biomedical R&D expenditures, *New England Journal of Medicine* 370: January 2014, 3–6.
7. Michael A. Heller et al. Can patents deter innovation? The anticommons in biomedical research, *Science* May 1: 1998, 701.
8. Jonathan R. Greene. Biochemistry sequence of the bacteriophage SPOI gene coding for transcription factor 1, a viral homologue of the bacterial type II DNA-binding proteins (HU protein/prokaryotic chromatin/transcription), *Proceedings of the National Academy of Sciences of the United States of America* 81: November 1984, 7031–7035.
9. Ibid.
10. *Myriad v. AMP.*
11. Patent US 20130317081—SERPINC1 iRNA.
12. *Myriad v. AMP.*
13. *Myriad v. AMP.*
14. Leahy–Smith America Invents Act Implementation | USPTO.
15. GeneDx, Inc. and its Parent BioReference Labs Announce http://www.prnewswire.com/news-releases/genedx-inc-and-its-parent-bio-reference-labs-announce-settlement-of-their-cancer-gene-testing-patent-dispute-with-myriad-genetics-300036231.html
16. Robert E. Yoches. Designing around patents, *China IP News* August: 2010.
17. Bridget M. Kuehn. Supreme court rules against gene patents, *JAMA* 310(4): July 2013, 357–359.
18. Nicholas Landau. The real impact for healthcare and biotechnology of the supreme court's decision in Myriad genetics, *Mondaq* July 24: 2013.
19. G. Steven Burrill et al. Biotech Industry Shatters Fundraising Records in 2014, *The Burrill Annual Report*, 2014.
20. Ibid.

21. Robert Cook-Deegan. Gene patents. ed. Mary Crowley, *From Birth to Death and Bench to Clinic: The Hastings Center Bioethics Briefing Book for Journalists, Policymakers, and Campaigns* (The Hastings Center; Garrison, NY, 2008): pp 69–72.
22. Emma Barraclough. What Myriad means for biotech, *WIPO Magazine* August: 2013.
23. Ibid.
24. Ibid.
25. J. Reichert. Biopharmaceutical approvals in the U.S. increase, *Regulatory Affairs Journal Pharma* July 1–7: 2004; see also Department of Health and Human Services Secretary's Advisory Committee on Genetics, Health, and Society. *Realizing the Potential of Pharmacogenomics: Opportunities and Challenges* (Department of Health and Human Services Secretary's Advisory Committee on Genetics, Health, and Society; Washington, DC, 2008): p. 11.
26. Ewen Callaway. Deal done on HeLa cell line, *Nature* 500: August 8, 2013, 133.

SECTION IV
Gene Modifications and Bioethics

6

Bioengineering and the Idea of Precision Medicine

Recent publications in scientific and philosophical journals, and popular newspapers such as *The New England Journal of Medicine, The Journal of the American Medical Association, The New York Times, The Washington Post,* and others have highlighted some of the current innovations in recombinant biotechnology. Of particular interest is the ability of researchers to manipulate DNA, genes, or entire genomes by studying the molecular pathways of a cluster of genes. It is not uncommon to hear that "scientists are designing human beings in the lab," "the era of designer babies has finally come," and perhaps sounding quite surreptitiously, "Eugenics is lurking," while others appear more gracious, "The era of precision medicine is finally here."[1] As Professor Jennifer Doudna has noted, "The idea that we can modify primates easily with this technology is powerful."[2] As with any new or scientific innovations, the reaction varies—while some may perceive the new corpus of knowledge in positive terms, others may be ambivalent, still others may be cautious in their perceptions. But why are people concerned about gene modification technologies? What are the applications of gene editing, and of what clinical value and prospects do they hold? The phenomenon of gene editing is opening new prospects and paradigms in molecular biology.[3] Indeed, it is bringing about some kind of epistemic or paradigm shift toward precision medicine, the redesign of species such as primates and changes in entire genomes. Despite the controversies surrounding these gene editing methods, proponents have taunted the benefits of the process. This chapter explores these technologies, the challenges they pose, as well as the ethical and policy ramifications for healthcare delivery and PM in particular.

What is gene editing?

As indicated in Chapter 2 of this book, there are at least 3000 genes responsible for mutations, insertions, deletions, or in scientific terms SNPs in the human genome (estimated to be at least three billion genes). Recombinant biotechnologies used for the sequencing of the human genome have become less costly and accessible to the ordinary scientist. As a result, scientists over the past few years have assiduously "targeted" modifications of the genome or specific gene of interest in which DNA is *inserted, replaced,* or removed from a genome using artificially engineered *nucleases.*[4] Nucleases are molecular scissors that are used to create double-stranded break in a genome or gene of interest in order for *exogenous* gene of interest to be introduced with the indigenous DNA's ability to *repair* and *recombine* the new ones. This process is known as *gene editing* in the genre of molecular biology.[5] For example, it is an undisputed fact that successful gene editing will ultimately result in a change in the genome of the patient. This is because during the "repair" process, the product or the *new gene* that emanates from the editing becomes incorporated or recombined into the genome and becomes inheritable. Gene editing could also be carried out on cluster of genes for desirable clinical significance such as sickle cell anemia (SCA) or enhancing a particular gene function for the "new gene (s)" to become an inheritable material. Other terms are gene corrections, gene modifications, gene additions/deletion, and gene manipulations or genomic bioengineering.

Characteristics of gene editing tools

Restriction enzymes are naturally found in bacteria where they function as defensive apparatus against viral infections or foreign DNA by cutting them.[6] Restriction enzymes are also known nucleases because they split or cleave DNA either at recognition sites or specific sites of the host bacteria. Because of their ability to "break," "cut," or "open" nucleic acids, they are often referred to as molecular scissors and have been significant in recombinant biotechnology. The cut may be a spontaneous single-stranded break (SSB) in a DNA sequence or protein, which could increase genetic instability and sometimes cause neurodegenerative diseases such as *Spinocerebellar Ataxia*. A SSB may also occur due to intracellular metabolites such as reactive oxygen. While these SSB occur frequently, the breaks are easily repaired by the DNA repair pathways. The cut may also be a double-stranded break (DSB) in the DNA sequence or protein which unlike SSB is rare. It is worth noting that DNAs have a natural or endogenous modification mechanism known as *restriction modification system* (RMS) that "repairs" damages or the molecular lesions internally. Once the repair mechanism is *activated*, it can result in either a homologous or nonhomologous recombination depending on the type of break in the DNA sequence. But the repair can also be done by bioengineering specific nucleases to repair the scar or the cleavage sites. In *homologous* repair, similar nucleotides or DNA molecules are exchanged or recombined in a double-strand break. This corrects aberrations or potentially harmful damages done to some genes during the process of cell division (meiosis). The nonhomologous mechanism also repairs DSB but does not use similar template during the repair process and the coded gene becomes nonfunctional among others. The process of gene editing involves the construction of nucleases that introduces DNA DSB at specific sites within a genome or gene of interest. This break or cut allows for gene editing technologies to be introduced: to turn off specific genes resulting in their loss of function in cells for therapeutic purposes or in some situations new copies of genes may be added for specific reasons such as bolstering the functions of a cluster of genes or individual genes.[7]

In brief, gene editing involves the recognition of specific gene sequence and the introduction of a restriction enzyme or endonuclease to cut resulting in a cleavage at the recognition sites typically leaving a lesion behind. A new DNA sequence or a carefully engineered genomic material may be introduced and once these bind properly, the DNA endogenously repairs and recombines the new strand into the genomes.[8]

Types of gene modification biotechnologies

Currently, there are four forms of gene editing biotechnologies or methods in use. They are

1. Clustered regularly interspaced short palindromic repeats (CRISPR) Cas9
2. Zinc finger nucleases (ZFN)[9]
3. Transcription activator-like effector nucleases (TALENs)
4. Meganuclease-reengineered homing endonucleases

The first and perhaps most popular system is known as CRISPR/Cas system. CRISPR is a bacterial adaptive system. Through the process of evolution, bacteria have developed this defensive mechanism against phage infections. When a virus infects bacteria, the CRISPR system incorporates foreign genomic materials and this becomes inheritable. Thus, the bacteria is able to develop immunity in future infections by recognizing the specific sites of the new infections and eliminate them at these recognition sites. The Cas9, in particular, can be used to cut genes at any loci within the genome or alter specific genes responsible for a particular pathway or a gene of interests. Currently, because of its versatility, the CRISPR Cas9 system has become one of the most popular molecular tools in bioengineering and specifically, gene modifications.[10]

The ZFN and TALEN are nucleases that recognize DNA domains in a genome.[11] The ZFN is protein-binding nuclease that breaks or cuts double-stranded DNA leading to a kind of site-specific mutagenesis. The cut leads to an activation of the DNA natural repair restriction modification mechanism (RM) to join the cuts by homologous and nonhomologous end joining (NHEJ) and this fixes the break. ZFNs could be introduced into embryonic cells by microinjection in anticipation of specific protein target for modifications such as deletions (knockout genes). Once the ZFNs reach their target and causes a break, NHEJ occurs leading to a *mismatch* of the DNA strand and loss of function of the genes since it has been eviscerated or spliced out.[12] It is important to note that this newly modified gene is inheritable and passed onto the next generation. TALENs have almost the same mechanism as ZFN. ZFN is currently being used in clinical trial to study the mechanism of HIV infection.[13] TALENs are made up of 33–34 amino acids fused with FokI nuclease to ensure specific cleavage of nucleotides.[14] Typically, TALEN induces a DSB in the DNA or gene of interest and inactivate them.

Classifications and applications of gene editing biotechnologies

Gene editing methods have a wide scope of applications in molecular biology due to their simplicity and precision. The technology may be classified according to many metrics or based on their specific utility. There are *reproductive* and *nonreproductive* applications of gene editing. Other classifications are therapeutic and nontherapeutic applications of gene modifications.

Generally, human reproduction is a significant facet for the survival of the species. The process of fertilization is also crucial since it confers viability to the sperm and the egg. But this process may be fraught with some challenges. Sometimes, couples or potential parents might have low sperm count (oligospermia), or low ovarian reserves (LOR), or, as a result of chemotherapy or other idiopathic reasons, are unable to experience natural fertilization; in some situations, either partner might carry known perilous genes. In both of these scenarios, gene editing technologies may be used to ameliorate the situation. It should be pointed out that beside these technologies, preimplantation genetic diagnosis and *in vitro* fertilization (IVF) gene modifications have already been used in assisted reproduction.[15]

In addition, in pre-fertilization reproductive contexts, typically, preimplementation genetic tests are conducted on the embryos or zygotes, which involve the removal of cells *in vitro* for genetic screening for traits. Diseases such as DMD and hemophilia are easily detectible through genetic testing. As a result, embryos manifesting these abnormalities may be discarded while the good ones are selected for implantations.[16] But editing technologies even go further than that. The genetic abnormalities could be detected by analyzing the genome of the sperm and egg donors before fertilization occur. The undesirable genetic traits may be deleted or corrected in order to prevent the clinical manifestations later in life. For example, mitochondrial DNA (mtDNA) mutations could manifest as diseases such as Lehigh syndrome, dystonia, and Pearson syndrome and these are serious debilitating clinical conditions. Currently, there are no known pharmacological or clinical treatments that could cure these diseases. Increasingly, IVF researchers and embryologists are using gene modification technologies such as TALEN or CRISPR Cas9 to alter nuclei of mtDNAs in order to curtail their transmission in the respective families. In a recent study, researchers noted "CRISPR-Cas9 technology, as well as other genome engineering methods, can be used to change the DNA in the nuclei of reproductive cells that transmit information from one generation to the next (an organism's 'germ line')."[17] Reddy et al. also conducted some studies using an animal model to investigate the application of the CRISPR Cas9 system.[18] It should be noted that some of these have been accepted for use in humans in some rare cases while a couple of clinical trials are taking place. For example, reported in a recent publication in *The New England Journal of Medicine,* Pablo Tebas and his astute team of investigators have genetically edited the autologous CD4 T cells of some patients with HIV and reintroduced the "modified cells" back to the patients with exceedingly high results with enormous ethical quagmires.[19] Furthermore, fertilization is critical in the process of reproduction in living beings especially among humans. The entire process of embryogenesis is critical for the health of any offspring, because genomic mutations in both or either parents could be transmitted to the offspring if not corrected. This is why it has been said that the health of the embryo is the health of a generation. Gene modifications are currently being explored as a tool to correct defective or mutated genes in the embryo. For example, CRISPR Cas9 has been microinjected into mouse embryos to correct *Crygc*gene responsible for cataracts.[20] This animal model has been successful in repairing the mutated genes. The gene editing could occur both *in vivo* and *in vitro.* In the case of infertility, the method could be used to correct genes responsible for this and, obviously, as Dr. Kathy Niakan indicated, to study embryogenesis in humans.

There are nonreproductive applications too where gene editing may be used *in situ* or *in vivo* for therapeutic and aesthetic purposes. With extensive knowledge of specific genes and their phenotypic manifestations, researchers could "customize" any type of gene of interest for clinical applications. A client could request for the bioengineering of muscles for agility for athletic purposes. A biologic could also be pharmacologically designed to induce mutated gene damage or block the functional manifestation of a harmful gene. Therapeutically, this could be the most useful and promising significance of gene editing in humans at any stage of the life cycle. Gene mutations occur at any level of human development due to many factors. Such mutations, as we have noted in Chapter 1 of this book, could be benign or could cause diseases such as cancer. Gene editing systems could be used in the deletions of mutated genes or the insertion of missing genes ostensibly geared toward the restoration to a normal genome. As will be discussed soon, gene editing seems to hold a promising key toward the reversal of HD, Alzheimer's, some oncogenes among others. Gene editing (GE) for therapeutic purposes may include germline modifications to bolster immunity (T-cell modifications) and for enhancements such as athletic performance.[21] In addition, germline modifications occur where the genome has been altered and obviously, the *new genome* becomes inheritable for future generations. This is by far one of the most controversial applications of gene engineering. As a diagnostic tool, some of the gene manipulative tools used to identify certain gene aberrations due to mutations or diseases they cause in humans. GE has been used to reengineer T-cells in order to bolster immunity in humans against viral infections, leukemia among others. Another cluster of applications are enhancements such as for sports and aesthetic purposes or designing specific bodily features such as transgenic eye colors, noses, and height. It should be noted here that animal models such as primates, pigs, and mice have already been accomplished. These discussions may be summed up in the following table:

1. Reproductive	a. Prefertilization b. Fertilization and reproductive purposes i. *In vivo*: embryos could genetically be altered ii. *In vitro:* preimplantation editing of embryos or nuclear transfers iii. Assisted reproduction**:** antidote to infertility treatment: low sperm or infertility in both sexes c. To study the process of embryogenesis and the various pathways
2. Nonreproductive	a. Therapeutic applications: i. Germline modifications ii. As a diagnostic tool iii. Immunobiotechnological application to bolster immunity b. Enhancements c. Aesthetics

Are these biotechnologies safe both in the short and long term, especially in humans? Why does the process of gene editing pique ethical concerns both among scientists and nonscientists without prejudice? Are these technologies intrinsically safe and precise as they are presumed and demonstrated within the genre of molecular biology? What are the limitations and potential effects? How does the clinical application of gene modification tools iterate the seeming old questions of eugenics? Could genome editing pose some challenges to the hitherto naturally endowed parental rights? These and other ethical and policy issues would be discussed in subsequent paragraphs.

Some ethical quagmires of gene modifications

Usually, when we hear of ethical concerns or discussions, they are often encapsulated in pejorative terms. That is to say, the negatives or what might be in contradistinction to the norm. In this context, however, by ethics, I mean both the good and the potential challenging issues in gene editing. This chapter will therefore explore the pros and cons of gene modification through the aperture of ethics in order to identify some potential common grounds. This is because there are obvious and indeed substantive therapeutic benefits accruing from gene editing.

According to a recent WHO report nearly 40 million people globally have HIV and AIDS.[22] HIV (human immune deficiency virus) is caused by a retroviral that affects the immune systems of the victims and progressively weakens them. This further leads to other "coinfections" such as tuberculosis and diarrhea. Over two million people die of these infections annually despite the availability of therapeutic interventions such as antiretroviral drugs to control the disease.[23] The clinical and socioeconomic impact of the disease have been significant especially in emerging economies, affecting some of their most productive workforce at epidemic proportions. Despite all current standard treatments, there are no specific cures that completely eviscerate the disease per se. But researchers have been assiduously exploring the plausibility of using gene editing as an alternative to cure HIV and AIDS as in the case of Timothy Brown.[24] He was known as the *Berlin Patient* and was born in Seattle, Washington. While studying in Germany, he was diagnosed with HIV and AIDS and acute myeloid leukemia (AML). After many fruitless treatments, his attending physicians recommended hematopoietic stem cell transplantation (stem cell from bone marrow) as a last resort. In 2007 and 2008, he received the transplant from a donor with a rare mutation of the "delta 32" (which normally retrogresses AIDS) found on the CCR5 genes.[25] Tests later showed that he was completely cured of the infection. But his case was obviously rare because the purpose for the transplant was for the leukemia but due to the inheritable changes of the genes in the transplant, he was clinically cured. The *Berlin Patient's* cure ultimately gave some impetus for further research and clinical applications of gene editing technologies to bolster the immune systems of other patients and to develop specific pharmacogenetically based biologics to cure some infections.

Researchers recently enrolled some HIV/AIDS patients in a clinical trial using one of the gene editing methods that targets the T cell's genes since HIV normally enters and disrupts the TD4 cells through the *CCR5* receptor leading up to AIDS. Using the ZNFs methods, researchers genetically modified the receptor gene in the patients' CCR5 genes. The CCR5-modified or edited CD4 T cells were then reintroduced into the patients. Subsequent tests showed significant increase in the CD4 T cells in the patients—an indication that the patient has developed immunity to the HIV infections. The researchers found that

> The gene-modified cells readily engrafted and persisted after adoptive transfer. Potential beneficial effects associated with the infusion of SB-728-T included increased levels of CD4 T cells. The observed relative survival advantage of the gene-modified cells during treatment interruption suggests that genome editing at the CCR5 locus confers a selective advantage to CD4 T cells in patients infected with HIV.[26]

These are obviously consistent with the ethical principle of *beneficence* because the health of the patient has been restored. We could deduce from these examples the onerous significance of the application of gene editing biotechnologies for many therapeutic purposes, in fact, even in public health and nutritional health. Gene editing could be explored to correct diseases caused by genetic aberrations such as phenylketonuria, cystic fibrosis, Tay–Sachs disease, oncogenes (that leads to cancer), and SAC among others. In addition, gene modification techniques could be beneficial to people with compromised or low immune systems or suffering from cancer.[27] Gene editing could be used to bioengineer personalized immune systems in humans in order to improve quality of life and curtail the incidence and potentials of infections leading up to diseases. The "potentials" of gene modifications in improving human life remains unprecedented and I believe the intent of researchers and clinical applications are inherently beneficent.

Another benefit is in the area of pharmaceutical development. The development of pharmaceuticals involves the "use" of both healthy patients and patients at various levels during clinical trials. Although clinical trials are generally considered safe, nonetheless the process may involve some risks and invasive procedures for enrollees (human subjects). The concept of beneficence implies some modicum of risk. The principle suggests that in an ethical quandary or muggy situation, "risk" ought to be carefully determined and apportioned against the backdrop of calculi of therapeutic success or benefits such that every member of society or subgroup shares the burden of the risks so that a beneficiary of a research is not overburdened. In other words, there is risk both anticipated and serendipitous even in research and invasive therapeutic procedures such as gene editing involving human subjects. Sometimes, clinical trials may go skewed with fatalities as with the case of Mr. Peter Munro, a 48-year-old, who died while taking Rhu Dex during a Phase I clinical trial involving the use of gene modification systems. Rhu Dex was purported to "...block[s] T-cell activation to prevent inflammation in rheumatoid arthritis...."[28] Obviously, this and others have posited serious setbacks in gene editing technologies for therapeutic purposes prompting many scholars calling on researchers to halt such "risky" clinical trials because the technology might not be safe and efficient. In fact, in a recent article in *JAMA* one of the original coinventors of the CRISPR Cas9 reiterated this:

> The lack of efficient, inexpensive, fast-to-design, and easy-to-use precision genetic tools has long been a limiting factor for the analysis of gene functions in animal models of human disease. Efficient genomic engineering to enable targeted genetic changes both in somatic cells and in the germlines of a wide variety of animals would facilitate pharmacological studies and the understanding of human disease pathways in ways not previously attainable. One of the most substantial types of genomic abnormality occurs due to rearrangement of non-homologous chromosome segments. In the past, modeling such chromosomal translocations in adult animals has been challenging due to the requirement for complex manipulation of DNA in germline cells. The CRISPR-Cas9 system enables induction of exact chromosomal translocations in somatic cells, thereby producing a much more robust and representative animal model of carcinogenesis.[29]

Despite these challenges enunciated above, some scholars are of the view that an "experimental use" of gene therapy be allowed although there are obvious undetermined risks. Proponents often adduced that where it is scientifically and, in particular, clinically established that every known therapeutic approach has been exhausted without a good result or cure, patients should be allowed to be treated with gene modification therapies and biologics on experimental and compassionate bases with approval from local IRBs. An experimental use shifts the axis of risks completely to the patient and family because if the therapy is not efficacious but causes some harm, then it will be in violations of the ethical principle of nonmaleficence (avoidance of harm). So, a careful and thorough clinical and prudential calculus has to be employed to insure that patients are prevented and insulated from any iota of perceived or actual harm. This model of balancing the risks calculi of using clinically unproven and potentially unsafe genomic therapy under the bedrock of beneficence and nonmaleficence came up recently in the case of Baby Layla Richards.[30] She was born healthy but her health deteriorated rapidly after a couple of weeks and she ended at the intensive unit of the hospital. She was diagnosed with acute lymphoblastic leukemia (ALL).[31] After all the usual treatments failed to cure her, her attending physicians at the Great Ormond Street Hospital suggested the experimental use of gene edited cells as an alternative and indeed as the last resort. As Layla's dad, Ashleigh noted, "It was scary to think that the treatment had never been used in a human before, but even with the risks there was no doubt that we wanted to try the treatment. She was sick and in lots of pain, so we had to do something."[32] The therapy for Layla comprised genetically bioengineered T cells with specific design to "target" leukemia or cancer cells. Upon the approval of the hospital's Institutional Review Board, she received the new cells and was clinically cured as confirmed by subsequent tests. And as a prominent oncologist, Alan Worsley commented upon hearing of Layla's recovery, "Re-engineering a patient's immune cells to target cancer has shown real promise in a small number of patients with leukaemia. This trial has adapted this treatment so that it's easier to make, and now we need to see if this new approach is effective. Finding a way to make this work in other types of cancer is the next big challenge."[33] Layla's prognosis and recovery are reminders of the potential that gene editing tools hold for precision medicine and the ethical trepidations they posit as well.

Globally, errors and fatalities and sometimes long-term effects of pharmaceuticals and invasive and noninvasive therapeutic procedures in clinical trials with human subjects are generally normative, even though less frequent, due to rigorous regulatory oversight and better *in vitro* and animal models. Gene editing has an important potential for developing better clinical models in ensuring safety and efficacy of drugs during Phases I and II. Gene alteration could potentially curtail invasive procedures during Phases I and II in human subjects typically in pharmacogenomics trials. Researchers may engineer genes *in vitro* or construct animal models that could be used for clinical trials especially when potentially dangerous pharmaceuticals are being tested, or develop vertebrae models (both *ex vivo* and *in vivo*) for clinical trials. That way, it has been surmised, the benefits to humans may be presumably maximized while the risks are passed on to the transgenic animals (which of course irk animal rights proponents as well). For example, Huang and his team genetically engineered monkeys using the CRISP/Cas9 technology by targeting certain genes.[34] This first transgenic primate has (as anticipated) generated a concatenation of debates with groups calling for immediate cessation of the researches involving the creations of further transgenic animals. Irrespective of the ethical quagmires raised, some researchers have already noted that transgenic primates could be used to study brain disorders that are generally difficult to model using *in vitro* or *in situ* or even *in silico*. That is, instead of enrolling human subjects, certain genes (known to have correlation with specific brain diseases) could be modified in transgenic animals and enroll them in Phase I clinical trial in order to limit the use of humans in clinical research, in general, involving some pharmaceuticals that could potentially alter the brain and the neural architecture of human research participants. Could the potential use of gene editing *ab initio* be consistent with the medical dictum, *primum non nocere* (above all, do no harm)? It is also worth noting that some scholars such as Moi Li and Christof von Kalle have raised some concerns about the plausibility of gene linkages into untargeted genes after gene insertions positing another safety concern.[35] These studies also suggest that some exogenous genes might float in untargeted cells in the genome of patients and these could potentially pose physiopathological risks.[36] Reporting on an experiment to create transgenic monkeys to study HD, Helen Chen observed that "viruses used to introduce the relevant gene had inserted extra copies randomly, intensifying the symptoms …" and five of the monkeys had to be euthanized because they clinically exhibited symptoms of the HD.[37] These further posit some epistemological and ethical questions. The question herein is the "absolute certainty" of the methodologies currently available. What are the unintended consequences? How stable are these new bioengineered genes introduced in the genome? Are these safely transmittable? Can new genes in terms of plasticity conform to genetic and evolutionary changes? As one of now, most of these questions remain contentious and it is in line with these that the WHO, the European Medicines Agency (EME), and the National Institutes of Health among others have not offered to fund these ventures involving human applications. Furthermore, there are platitudes of technical hiccups worth at this point. Even though the technology associated with gene alterations seems to be precise, pragmatically, it is always not the case. For example, pharmaceuticals or biologics may be successfully delivered to the targeted gene but not necessarily to the right cells. As one pharmaceutical administrator pointed out:

> Among the unresolved technical challenges is figuring out the best way to deliver the drug to the cells it is targeting. The RNA gets the drug to the right gene, but it doesn't guide the drug to the right cells. Researchers are exploring hitching a ride on a harmless virus called the adeno-associated virus, or on tiny lipid nano particles. Both potential carriers have been under study for delivery of other kinds of drugs. "We still need to identify the best delivery methods."[38]

Ultimately, transgenic animals for clinical use on the expediency of their assumed and theoretical bases are genuinely promising toward the redefinition of clinical trials and the next generation of protein-based pharmaceuticals and PM and, therapeutically, procedures such as personalized gene surgery.

Furthermore, successful gene editing will ultimately result in a change in the genome of the patient.[39] This is because during the "repair" process, the product or the new gene that emanates from the editing becomes incorporated and conserved into the genome. Gene editing could also be carried out on a cluster of genes for desirable clinical benefits. While this may potentially lead to clinical benefits such as curing a particular disease such as SCA or enhancing or bolstering a particular gene function to understand molecular pathway, the "new gene(s)" as it were becomes an *inheritable material.* A progeny of the patient will inherit this new gene now in the genome of the patient. Gene editing may be akin to the concept of designing future generations and their genome. That is to say, gene editing could interfere with normal biological and genetic process leading up to a new form of life; of which the future progeny has no hands or autonomic choices in. As mentioned earlier, SAC is debilitating disease but could also be advantageous for people living in malaria-prone areas. Therefore, editing the sickle cell gene could pose some clinical challenges for the recipient and the progenies who might be living in these areas. I believe that passionate debates guided by a deep sense of transparency, openness to accepting corrections from opponents, making data available to the public in a bid to arriving at a better acceptance of the good and the limits of the biotechnologies are obviously essential for the common good.

In addition, the mere perception that embryonic gene editing or gestational editing is feasible and have been performed heightens societal concerns of PM. Some diseases could be identified during embryogenesis and the entire gestational periods in primates and humans. Tests such as pregnancy-associated plasma protein screening (PAPP-A) and diagnostics for Leber congenital amaurosis (which causes blindness in children as young as a year old) could be conducted in early placental development. Abnormal levels of pappalysin-1 protein (encoded by the PAPPA gene) could be symptomatic of chromosomal abnormalities.[40] These are not clinically curable but manageable. Because of these challenges, currently, the focus has shifted on gene modification technologies as the last resort in order to potentially correct the genetic abnormalities. The editing process implies the incorporation of "exogenous genes" into the genome of embryos that may transmit these to their progenies. While these are still under consideration and are at exploratory stages (since most regulatory agencies have placed embargos or moratoriums on *in vivo* gene editing procedures), they seem to hold great prospects for precision medicine. But there are no guarantees that the new or the exogenous genes would remain stable. This is because there could be posttranslational modifications in which edited genes could change and potentially obliterate the initial purpose. Could the preponderance benefits outweigh the risks? Also, a new gamete from a third party (a donor) may be genetically edited for recipients during IVF procedures. There have been ethical debates on whether a child born of this procedure should be informed about his genetic parents. Currently, Britain has issued a directive pointing to disclosure. The Presidential Commission on Bioethics in the United States is currently exploring this. It is in view of this that the words of one of the coinventors of the CRISPR Cas9 system offered a somberly reflection worth pondering:

> The advent of CRISPR-Cas9 technology underscores the importance of basic research for advancing medicine. Once the molecular mechanism of CRISPR-Cas9 was understood, it could be harnessed for applications not previously imagined. Ongoing research focuses on determining Cas9-mediated gene editing specificity, as well as increasing the frequency of homology-directed DNA cleavage repair. Furthermore, methods for delivering Cas9 and its guide RNAs into cells need to be tested in disease-relevant tissues and animal models. The era of genome editing raises ethical questions that will need to be addressed by scientists and society at large. How should such a powerful tool be used to ensure maximum benefit while minimizing risks? It will be imperative that nonscientists understand the basics of this technology to facilitate rational public discourse. Regulatory agencies will also need to consider how best to foster responsible use of CRISPR-Cas9 technology without inhibiting appropriate research and development.[41]

I believe these words are consistent reminders of the prospects and the limits of scientific innovations. From the theoretical point of view, gene editing is novel and offers a glimpse of endless hope in the redefinition of science and in particular precision medicine. However, both in the scientific community, within regulatory agencies, advocacy groups and the general public there seem to be a consistent call on biomedical researchers to thread on the conduit of meticulous caution rather than full acceptance of the technologies. Rather than truncate or completely condemn these technologies, I believe an openness marked by the aura of mutual discussions devoid of surreptitiousness could facilitate the debate recognizing dissenting views even internationally. It seems we are at the verge of an important scientific revolution that is anticipated to shake the very epicenter of medicine and a unique perspective and understanding of living entities at the molecular levels.

In addition to the above, gene editing technologies could also play an important part in the issue of *aesthetics*. It is not uncommon to hear about "designer babies" or genetic enhancements. There seem to be an intrinsic nature of humans that points to a sense of imperfection. Human beings by nature are always working toward some form of perfection. As biological systems, humans are not perfect per se—people have desires to be a bit taller, athletic, and muscular, live without wrinkles, and have that perfect intelligent quotient (IQ) or the ability to excel in some facet of life. There is thus a deeply ingratiated inclination in humans to perpetually strive toward the attainment of an inexplicable perfection. These goals are exemplified in the perception of their lives and their environment and other human beings they interact with. The perception of aesthetics or beauty is ingrained in us to the extent that in our generation, we spend a fortune to "improve" or redo our physical features for specific purposes. People raise concerns about their heights, chick size, breast size, penile size, hair color, eye color, skin texture, wrinkles, and an avalanche of others. Pragmatically, the perception of beauty though seemingly subjective, transcends every known epoch. We love beautiful things or things that appear to us to be beautiful. In one of his seminal works, *The Metaphysics*, Aristotle noted in *Book Delta,* using the Greek word, *telos* to give some reflections and taxonomies of perfection. While *telos* typically translated as a goal or purpose, Aristotle explained that perfection

1. Which is complete—which contains all the requisite parts
2. Which is so good that nothing of the kind could be better
3. Which has attained its purpose

This tripartite exposition is worth noting and applicable to our reflections on "perfecting genes." Perfection occurs when something or an entity attains its "completeness." What constitutes gene perfection? Is there any human genome that is actually, and scientifically demonstrable to be pure devoid of any trace of mutation either spontaneous or through the process of evolution? Should a gamete or a person's gene be edited for aesthetic reasons to ensure that the person is complete and so good for his or her own goal? I believe this is a subjective question that an individual considering gene editing ought to ponder. As, Hume also indicated in his famous work *Moral and Political* (1742): "Beauty in things exists merely in the mind which contemplates them" and of course, Shakespeare once wrote: "Beauty is bought by judgement of the eye. Not utter'd by base sale of chapmen's tongues." These are important aspects of our neural architecture as human beings. Currently, gene editing technologies could be used to insert any desirable traits into the genome to alter genes responsible for height, eye color, or into the human gametes or embryos that may be of no therapeutic purposes other than aesthetics. However, gene editing has been used in assisted reproductive situations especially in the wake of infertility where cytoplasmic transfer in particular has been used by Cohen and his team to assist couples having trouble conceiving naturally. Cytoplasmic transfer involves a transfer of fertile eggs from a donor to infertile recipients through microinjection. mtDNA from the donor rejuvenates the recipients' eggs leading to successful gametes and embryonic developments that contain mtDNA of two mothers and the DNA of a father. Although no serious risks have been reported about the children born through this technique, the researchers, however, indicated that the procedure was still at experimental or research stages. CRISPR Cas9 could be used to "design" babies by editing genes responsible for certain phenotypes such as nose size for aesthetic goals without any therapeutic purposes. But this may be considered "ethical" because parents have reproductive rights such as the choice of their spouses or the person they desire to share reproductive privileges with, the time they desire to have kids, and the number of kids they want, and sometimes the sex of the baby. These are all important parental and reproductive privileges recognized by law and as indicated above are ethical and seemingly consistent with natural law. In addition, parents have the legal and ethical authority to make decisions in the interest of their children including the unborn. Gene editing poses a caveat and a modicum of debate on the plausibility of defining some scope of limitations. Should parents have the right to design their babies under the expediency of aesthetics? For example, if a parent "designed" his/her child to have blue eyes and as tall as 6"9 because the "parents" felt these traits may make him a better athlete, could the child undo these traits (even though not feasible now but possible given the technologies)? Will children have the legal rights to sue the parents for "designing" them for not allowing them to develop and grow "naturally"? If a child is genetically designed for a purpose but did not grow up to achieve that, could that lead to a law suit by the parents? WHAT IF, along the line, the inserted genes undergo some spontaneous changes (which is not uncommon) with debilitating results, who has the fiduciary obligation to offer care? These questions continue to emerge as the debates on gene editing unfolds. I am afraid responses to these may be assuredly hypothetical rather than pragmatic. I believe these quagmires among others might influence Doudna's quest for paucity in these words: "My colleagues and I felt that it was critical to initiate a public discussion of the appropriate use of this technology, and to call for a voluntary ban on human germline editing for clinical applications at the present time…".[42] But in a strongly worded dissent to the call for abrupt paucity GE research, Chris Gyngell notes

> Far from being wrong, the research by Huang and colleagues is ethically imperative. Such research not only has the potential to provide permanent cures for genetic diseases, it also holds the potential to correct the genetic contribution to common diseases like diabetes. It even has the potential to give people the capacity to age better—some extremely people age well into 90 s and 100 s. Age-related disease alone kills around 30 million people per year.[43]

Gyngell suggests that the ethical arguments undergirding the ban on GE research are fundamentally flawed. With a sense of optimism, Gyngell calls for an openness and continual support for gene modification research asserting that the ban could stall potentially valuable therapeutic research and data rhapsodized in this rhetorical question: *Imagine that I am a scientist. I have a promising candidate treatment that could save the lives of 30 million people per year. I decide not to continue the research. I am responsible for the death of those 30 million people if my research would have led to a cure.*[44] This was in the contexts of *Nature* and scientific journals' decisions not to publish Huang and colleagues' initial embryonic gene editing research paper.

In brief, gene editing biotechnologies such as ZFNs, CRISPR Cas9, and TALEN undoubtedly hold myriads of prospects in precision. As with any new scientific innovations, sociological, ethical, legal, and policy challenges would continue to emerge because by its nature is also a social enterprise. Responses and policies emanating from these debates will chart the contours and the course with which these novel innovations will impact humanity as a whole. I believe some of the ethical issues discussed here will become some of the litmus tests for the acceptability and integration into mainstream clinical care and biomedical research.

End notes

1. Jacob J. National academies to establish human gene editing guidelines, *JAMA* 314(4): 2015, 330; T. Hampton. Gene researchers work to engineer HIV-resistant cells, *JAMA* 312(4): 2014, 323–325; The Gene. The biotech century: Harnessing the gene and remaking the world. *JAMA* 280(6): 1998; T. Hampton. New method allows for precise gene targeting in monkeys. *JAMA* 311(9): 2014, 894; H.T. Greely. Genetic modification. *JAMA* 292(11): 2004, 1374–1375; Eric B. Kmiec. Is the age of genetic surgery finally upon us? *Surgical Oncology* 24(2): June 2015, 95–99; Pablo Tebas et al. Gene editing of CCR5 in autologous CD4 T cells of persons infected with HIV, *New England Journal of Medicine* 370(10): March 6, 2014; Toni Cathomen, PhD et al. Translating the genomic revolution—Targeted genome editing in primates, *New England Journal of Medicine* 370(24): June 14, 2014; see also Tanya Lewis 2 leading biologists say we should allow gene editing on human embryos, *Business Insider Australia* November 25: 2015; Jeffry D Sander et al. CRISPR-Cas systems for editing, regulating and targeting genomes, *Nature Biotechnology* 32: 2014, 347–355; Elizabeth Alter. The risks of assisting evolution, *New York Times* November 10: 2015; P.R. Reilly. Eugenics and involuntary sterilization: 1907–2015, *Annual Review of Genomics and Human Genetics* 16: 2015, 351–368.
2. http://www.technologyreview.com/featuredstory/526511/genome-editing/; see also, Amanda Schaffer. Gene editing: The impact, *MIT Review* April 23: 2014.
3. Heidi Ledford. CRISPR, the disruptor, *Nature* June 3: 2015.
4. Sara Reardon. US science academies take on human-genome editing, *Nature* May 18: 2015; Jeantine Lunshof. Regulate gene editing in wild animals, *Nature* 521(127): May 12, 2015; Heidi Ledford. Embryo editing sparks epic debate, *Nature* April: 2015; Sarah Reardon et al. Chinese scientists genetically modify human embryos, *Nature* April 22: 2015
5. Heidi Ledford. Mini enzyme moves gene editing closer to the clinic, *Nature* April 1: 2015; T. Hampton. *New Method Allows for Precise Gene Targeting in Monkeys*; H.T. Greely. *Genetic Modification*; Eric B. Kmiec. *Is the Age of Genetic Surgery Finally Upon Us?*
6. Doudna J. et al. (eds). *The Use of CRISPR/cas9, ZFNs, TALENs in Generating Site Specific Genome Alterations in* Methods in Enzymology Vol 546 (Academic Press; New York, November 4, 2014); P. Mali et al. RNA-guided human genome engineering via Cas9, Science 339: 2013, 823–826. See also, L. Bazaar. *Methods in*

Biotechnology Lecture Notes, Georgetown University, 2012; L. Cong et al. Multiplex genome engineering using CRISPR/Cas systems, *Science* 339: 2013, 819–823; F. Zhang et al. Efficient construction of sequence-specific TAL effectors for modulating mammalian transcription, *Nature Biotechnology* 29: 2011, 149–153.

7. Ibid. Doudna J. et al. (eds). *The Use of CRISPR/cas9, ZFNs, TALENs in Generating Site Specific Genome Alterations in* Methods in Enzymology Vol 546.
8. Ibid.
9. E. Kim et al. Precision genome engineering with programmable DNA-nicking enzymes, *Genome Research* 22: 2012, 1327–1333; C.L. Ramirez et al. Engineered zinc finger nickases induce homology-directed repair with reduced mutagenic effects, *Nucleic Acids Research* 40: 2012, 5560–5568; J. Wang et al. Targeted gene addition to a predetermined site in the human genome using a ZFN-based nicking enzyme, *Genome Research* 22: 2012, 1316–1326.
10. Ibid. See also Ignazio Maggio et al. Genome editing at the crossroads of delivery, specificity, and fidelity, *Trends in Biotechnology* 33(5): May 2015, 280–291.
11. F.D. Urnov et al. Genome editing with engineered zinc finger nucleases, *Nature Reviews Genetics*, 11: 2010, 636–646.
12. Ibid.
13. N. Holt. *Nature Biotechnology* 28: 2010, 839–847.
14. (Deng et al., 2014).
15. Marcy Darnovsky. A slippery slope to human germline modification. See also Y. Verlinsky et al. Preimplantation diagnosis for Fanconi anemia combined with HLA matching, *JAMA* 285(24): 2001, 3130–3133.
16. This poses some ethical discussions beyond the scope of this book.
17. Jennifer Doudna. A prudent path forward for genomic engineering and germline gene modification, *Science* April 3: 2015.
18. Reddy et al. Selective elimination of mitochondrial mutations in the germline by genome editing.
19. Pablo Tebas, MD. Gene editing of CCR5 in autologous CD4 T cells of persons infected with HIV, *Nature*.
20. Yuxuan Wu et al. Correction of a genetic disease in mouse via use of CRISPR-Cas9, *Cell Stem Cell* 13(65): December 2013, 659–662.
21. Ibid.
22. WHO | HIV/AIDS—World Health Organization.
23. Ibid.
24. Gero Hütter et al. Long-term control of HIV by CCR5 Delta32/Delta32 stem-cell transplantation, *New England Journal of Medicine* 360: 2009, 692–698. See also, J.A. Levy. Not an HIV cure, but encouraging new directions, *New England Journal of Medicine* 360: February 12, 2009, 724.
25. J. Stephenson. Gene mutation link with HIV resistance, *JAMA* 286(12): 2001, 1441–1442.
26. Pablo Tebas. Gene editing of *CCR5* in autologous CD4 T cells of persons infected with HIV, *New England Journal of Medicine* March: 2014.

27. S.R. Husain et al. Gene therapy for cancer: Regulatory considerations for approval, *Cancer Gene Therapy* 2015.
28. Megan Scudellari, Subject death halts clinical trial, *Scientists* July 8: 2008.
29. Jennifer A. Doudna, PhD. Genomic engineering and the future of medicine, *JAMA* 313(8): February 24, 2015.
30. Karen Zusi. Gene editing treats leukemia, *The Scientists* November 6: 2015; see also Ian Sample, Baby girl is first in the world to be treated with 'designer ..., *The Guardian* November 12: 2015.
31. Ibid.
32. Ibid.
33. Ibid.
34. Helen Shen. First monkeys with customized mutations born, *Nature* January 30: 2014.
35. Hacein-Bey-Abina et al. *LMO2*-associated clonal T cell proliferation in two patients after gene therapy for SCID-X1, *Science* 302(5644): October 17, 2003, 415–419.
36. Ibid.
37. Helen Chen. Precision gene editing paves way for transgenic monkeys, *Nature* 503(14): November 7, 2013.
38. http://www.wsj.com/articles/why-gene-editing-technology-has-scientists-excited-1434985998.
39. Moi Li et al. A cut above the rest: Targeted genome editing technologies in human pluripotent stem cells, *Journal of Biological Chemistry* February 21: 2014
40. A.E. Bolton et al. Measurement of the pregnancy-associated proteins, placental protein 14 and pregnancy-associated plasma protein A in human seminal plasma, *Clinical Reproduction and Fertility* 4(3): 1986, 233–240.
41. Jennifer A. Doudna, *JAMA* March: 2015
42. *Doudna in* Ethical and regulatory reflections on CRISPR gene editing revolution. Genetic Literacy Project (Jon Entine, June 25, 2015).
43. Chris Gyngell et al. The moral imperative to research editing embryos: The need to modify Nature and Science, *Practical Ethics* April 23: 2015.
44. Ibid.

7

Policy, Bioethics, and Bioengineering

Scientific discoveries occur within a *sitz-im-leben* (a social–cultural context) and therefore constitute a social phenomenon. At the micro level, scientific discoveries and innovations occur within their respective communities and specialties, and within the larger social contexts. Validation of the process within these social contexts not only authenticates the novelties of the discoveries but also legitimizes them for the entire society because new theories and discoveries have potentials or direct benefits and application to society. This often leads to some paradigm shifts at the micro and macro levels, depending on how people *react* and *respond* to the new discovery or its paradigmatic technological applications in resolving some enigma in society. This is because society has some vested interests in science since scientific and technological innovations seem to define the advancements or the enervation of society. But not all discoveries have been accepted with the same fervor and vim, partly due to some precedents, which are beyond the scope of this chapter. Science is a dynamic enterprise that is inexplicably open to constant change; either improving previous discoveries, new insights, or discarding discoveries due to error, new paradigms, or some kind of axiomatic or epistemic shifts among other reasons.[1]

Furthermore, the process of scientific innovations especially involving human subjects and animals are increasingly scrutinized with heightened regulatory and policy apertures due to substantive and documented exploitation of vulnerable populations. As a result, genomic innovations and researches are often perceived with some sense of intellectual skepticism and ambivalence. Such reaction from society may sometimes be subjectively emotive—it is difficult to quantify how individuals and societal responses are. Nonetheless, these reactions may often translate into policies to guide society make objective and socio-ethical policies with the bid to protecting society, in general, and in particular, vulnerable populations who may be recipients or users of these new scientific discoveries and technologies.

At the emergence of recombinant biotechnologies in the 1970s, scientists and policymakers raised concerns and wanted to deliberate further on the safety of this new scientific venture. This led to the first meeting to deliberate specifically on gene splicing and recombinant methods such as cloning, genetically modified organisms in general (including plants and animals), the emergence of reproductive fertilization technologies among others in the famous Asilomar Conference on Recombinant DNA in 1975 to discuss the regulatory parameters and the issue of risks and safety. It is worth noting that even though the meeting was at the behest of scientists, the 140 stalwart attendees were professionally diverse, including legal experts, physicians, and biologists, to deliberate on the prospects of the genre of DNA recombinant science and to craft some guidance and policies to ensure that research and the new impetus in molecular biology proceeded "cautiously." As the introductory part of the proceedings of the Asilomar Conference noted

> This meeting was organized to review scientific progress in research on recombinant DNA molecules and to discuss appropriate ways to deal with the potential biohazards of this work. Impressive scientific achievements have already been made in this field and these techniques have a remarkable potential for furthering our understanding of fundamental biochemical processes in pro- and eukaryotic cells. The use of recombinant DNA methodology promises to revolutionize the practice of molecular biology. While there has as yet been no practical application of the new techniques, there is every reason to believe that they will have significant practical utility in the future.

The Asilomar Conference discussed several issues particularly on the potential risks of recombinant biotechnologies involving the use of microorganisms such as bacteriophages, *Escherichia coli,* and others in manipulating DNAs among others. This conference, still considered to be one of great significance, in recent memories seems to be a pacesetter for most current scientific conferences to clarify contentious issues. It is not surprising to hear of scientists and policymakers using the expression, an "Asilomar moment" when in doubt or in a conundrum about some controversial and new scientific innovations that might be in dire need of some form of deliberations! In view of this and many other scientific conferences, many guidelines and documents are issued to guide further research or dissuade researchers from undertaking certain researches until there is a modicum of clarity and certainty within the scientific community. As noted above, responses may vary depending on many factors and may tacitly influence how the scientific community may proceed. In view of these, three noticeable approaches have emerged in light of the controversies surrounding gene modifications in biomedical research viz.

1. The precautionary approach
2. The moratoria approach
3. The noninterventionists approach

It is worth noting that these approaches may overlap for a particular scientific controversy or conundrum even within the same scientific community or geopolitical and social contexts. In the following paragraphs, I will expatiate on these approaches with the requisite dexterity.

Precautionary approach

What is this approach and why is it applicable to our topic of discussion? This approach emanates from the *precautionary principle* typically used in international policy and regulatory studies.[2] According to the dictionary, *precaution* is "an action taken to protect against possible harm or trouble or to limit the damage if something goes wrong." Precautionary actions are typically preemptive or preventative measures based on some calculi of risks in anticipating a modicum of result or (in the context of gene editing) an outcome that protects human life in general and patients in particular. Precautionary measures are taken under the aegis of uncertainty or indeterminate consequences of a situation. The precautionary approach wobbles on some basic principles that in the absence of consensus among experts, herein scientists, on an issue such as biomedical research that has the potentials of affecting society and human beings in particular, the experts must proceed with caution and the burden of risks shifts to them rather than the public. As Van Alles et al. noted, "The precautionary principle legitimates decisions and actions in situations characterised by uncertainty. It is generally agreed that uncertainty is the essence of the precautionary principle .... Both risk and uncertainty are thus central notions in the whole precautionary endeavour, both in terms of decision-making and decision-support."[3] This approach may be applicable in the context of the debate on gene editing. There is growing number of scholars and opinion leaders and even some regulatory agencies vying for a precautionary approach in addressing the quagmires posited by gene modifications on the basis that the technology might not be safe, and there seem to be many unanswered technical questions as well as genuine concerns about the potential or actual harm that may induce a precautionary approach to addressing the debate in gene engineering. The precautionary approach is increasingly popular within the scientific community most notably in Europe and now almost globally. The precautionary approach does not call for a complete ban on biomedical researches using the gene modification systems; rather it calls for reasonable and prudent use to mitigate risks. As noted above, scientific discoveries follow a process. The processes typically unfold gradually with increasing clarity and certainty from within the scientific community and society in general. The process may generate copious disagreements among experts due to conceptual, data, factual discrepancies, and misunderstandings. Sometimes, the scientific method may be flawed or some of the discoveries could not be validated or may be deemed too risky for replications. These in turn may generate doubts and as the aphorism goes *in dubio, abstine* (when in doubt, abstain). Gene editing is undoubtedly at its neophytic stages of development. The amount of researches or scientific papers written so far seems skeletal compared to gene sequencing or other scientific novelties. More so, there are many technical hiccups coterminous with the technologies. In a recent article, De Souza offers a cautionary reflection:

> But the potential of these tools will not be fully realized until they can be simply and efficiently designed (or inexpensively purchased) by any laboratory and unless they can achieve the desired editing outcome at essentially any target sequence. Methods development in this area thus continues at a fast pace. ...Continuing progress on these and other fronts should hopefully put flexible and effective genome-editing tools into the hands of many scientists in the near future.[4]

The CRISP Cas9 and other GE methods are not perfect per se and indeed posit a cluster of technical hurdles. Sometimes, GE may result in mismatches or the exogenous DNA might have missed the targeted sequence of interest. More so, there are concerns of posttranslational modifications or protein stability. Another concern is bioequivalence. Is that how the equivalent edited gene is going to manifest in the genome? As Mali and Church also noted in a comparative study, "An increasingly recognized constraint limiting Cas9-mediated genome engineering applications concerns their specificity of targeting. The sgRNA-Cas9 complexes are in general tolerant of 1–3 mismatches in their target and occasionally more, with the actual specificity being a function of the Cas9 ortholog, the sgRNA architecture, the targeted sequence, the PAM, and also the relative dose and duration of these reagents."[5] Indeed one of the prominent scientists and astute users of the CRISP Cas9 accentuated these challenges associated with the method in these words:

> Two issues remain outstanding: evaluating and reducing off-target effects. A number of studies have attempted to evaluate the targeting specificities of ZFN, TALEN and Cas9 nucleases. The limited number of studies characterizing ZFN and TALENs specificity have only highlighted the challenges of detecting ZFN and TALEN off-target activity. Of note, the two independent studies attempting to characterize the off-target profile of the same pair of CCR5-targeting ZFNs have returned distinct and non-overlapping lists of off-target sites, which highlights the challenges associated with analysis of nuclease specificity.[6]

Thus, the potential impact and risks of off-targets effects on edited genes or genomes remain uncertain. It thus seems or suffices to say that the technologies still work in progress contrary to how they have been presented as precise, self-contained, and accurate. It is in view of these that some ethicists in particular have called for cautionary approach and an "Asilomar moment" ostensibly to ponder, reflect, and formulate some guidance on the way forward. One of such pedigree in ethics, Arthur Caplan, puts it vividly, "No one should be doing anything right now until we figure out what the hell is going on with this technique in animals. That's how we perfected in vitro fertilization, and that's how you establish safety—you do it in animals first."[7]

Another gray area oscillates on *verification* and *replication* of the methods in therapeutic applications. For example, when Huang et al. bioengineered the transgenic monkeys for HD, it was reported that some of the primates developed the disease to the extent that they were euthanized while others survive. These and other examples seem to substantiate the cautionary or the prudential approach on the use of gene editing biotechnologies especially in therapeutic applications.

Another perspective of the precautionary approach is the thesis for the partial ban or selective use of the GE biotechnologies. This approach obviously acknowledges the potentials these innovations hold especially their therapeutic imports. But as already intimated above, GE posits a labyrinth of challenges. In addition to the intellectual challenges, there are value-based or socio-ethical challenges worth noting. To assuage these, the United Kingdom currently allows the GE in animal models both *in vivo* and *ex vivo* within the aperture of their regulatory frameworks. To that effect, gene editing is generally allowed in animals but restricted in humans. Human gene of targets may be edited as far it is *ex vivo* in the laboratory but cannot be performed on humans *in vivo*. Consequently, clinical research on human genome is also restricted. It is also worth pointing out that despite restrictions on gene editing on humans, there is a growing call for the lifting of the ban on research on human embryos. Just at the threshold of completing this manuscript, on September 15, 2015, Kathy Niakan, a renowned embryologist at the Francis Crick Institute had applied to Human Fertilisation and Embryology Authority (HFEA) based in London for license to use gene modification biotechnologies to study and

> To provide further fundamental insights into early human development we are proposing to test the function of genes using gene editing and transfection approaches that are currently permitted under the HFE Act 2008 We also propose to use new methods based on CRIPSR/Cas9 which allows very specific alterations to be made to the genome. By applying more precise and efficient methods in our research we hope to require fewer embryos and be more successful than the other methods currently used.[8]

Dr. Niakan argues persuasively that "... in line with HFEA regulations, any donated embryos would be used for research purposes only. These embryos would be donated by informed consent and surplus to IVF treatment."[9] She seems to recognize the general precautionary approach within the scientific community as well as social concerns about the use of human embryos. But would IC preclude the fact that the embryos may be of human origin? Assuming the license is granted, what will happen to the "rest of the embryos" after the research is conducted? In a tersely worded rebuttal, the Director of the NIH, said among others "...the concept of altering the human germline in embryos for clinical purposes ... has been viewed almost universally as a line that should not be crossed."[10] In other words, for the Director of the NIH, there can be no *a priori* or *prima facie* philosophical, scientific, policy or legal grounds for the use of human embryos especially in the context of gene editing. This is not to dissuade scientists from pursuing applications of gene editing in other areas not involving human embryos. Nonetheless, these significant questions await the decisions of the HFEA whether to grant or deny the license.

As a result of the above discussion, some scholars have called for a *total ban* on the use of gene editing biotechnologies in human biomedical researches until the nutty gritty of the challenges are clearly and demonstratively established to mitigate potential and actual risks in the short and long-term impacts on humans.[11] Increasingly, the anomalies seem to bring about some kind of "crises in science" because whereas most scientists would have wanted to proceed with the *status quo,* society in general is increasingly issuing caution. Currently, in Europe, North America, and others, there are severe restrictions on gene editing in humans especially human embryos and primates. In brief, an awareness of these anomalies often led to a "crises in science"—the effect of such discoveries seems to last for a period of time and sometimes transcends generations. Because such anomalies as Kuhn notes often led to dislodging of significant paradigms or paradigm shifts in scientific theories—this new paradigm may not be accepted easily within the scientific community and I believe the current crises oscillating on gene editing is reflective of this *Kuhntian* assertion![12] In brief, scientific theories and innovations sometimes lead to crises in science when new methods or processes challenge existing ones that leads to paradigm shifts in science. As a consequence of the above, two approaches could be synthesized from the "precautionary approach." The first is the call for "Moratorium" or total ban on the use of gene editing techniques in humans especially human embryos while a *noninterventionist approach* or the *proactionary approach* seem to precipitate from the debates.

Moratoria approach

Generally, a moratorium is used in resolving conflicts, sometimes to enable feuding or parties pause to harsh out differences in order to agree on term or permanent solutions in the interest of peace or other fiduciary, economic, political, and policy reasons. Moratoriums in biomedical and allied health sciences have been used in professional meetings or in academic fields when uncertainty or some discrepancies occur that might be detrimental to the health of the public, patients, or the genre of science in general.[13] From a definitional perspective, a moratorium is a *formally agreed period* during which an activity is halted or a planned activity is postponed ostensibly due to unanimous well-grounded concerns. One of the turning points in the discourse on gene editing came in an article in the *Journal of Protein and Cell* captioned: "CRISPR/Cas9-Mediated Gene Editing in Human Tripronuclear Zygotes" and later digressed in *Nature* as "Chinese Scientists Genetically Modify Human Embryos." The articles detailed the processes by which Chinese researchers (Liang et al.) genetically modified human embryos using CRISPR Cas9 systems. These were followed by a medley of publications with provocative titles such as "Ethics of Embryo Editing Paper Divides Scientists" and "US Science Academies Take on Human-Genome Editing" among others. One proliferating theme that is fueling the debate is the mere suggestion that edited human germline or genes could be inherited and potentially transmitted to another generation. As a consequence, some proponents of the moratorium approach argue that gene editing research should be truncated.[14] A synthesis of the arguments so far does not offer any conclusive and substantive scientific evidence that any of the gene editing tools poses any clinical risks. Nonetheless, there have been calls both within the academic scientific community, policymakers, value-based groups, and funding agencies such as the NIH for a halt to all gene editing procedures on human embryos.

The first objection to human embryonic research oscillates on the notion of uncertainty and impact of altered genes on future generations. I postulate this as the "gate-keeper" argument. But what does the gate-keeper argument entail? The argument suggests current generations have the onerous and fiduciary responsibility to ensure that the gene pools are not changed or altered for nontherapeutic reasons as encapsulated in the words of the Director of the NIH, Dr. Francis Collins's suggestions that modifications in embryonic genes "affect the next generation without their consent." There are two elements in this statement worth expatiating. First, the gate-keeper argument is premised on the protection of the future generation since their very existence is inexplicably linked with the quality of life of the current generation. It is a case of natural justice that vulnerable populations are protected. Obviously, the unborn or the next generation will inherit genes from current populations: it makes sense (at least, logically) to ensure that any potential threats to them are truncated if not completely eliminated. This argument is supported by basic science and some precedence, for example, exposure of DNA to chemicals during methylation is known to alter some genes, which may be transmitted to progenies or cause some epigenetic changes.[15] These changes could have severe consequences for those who inherit these altered genes. It is in view of these that some ethicists are pushing the alarm button that "genome editing in human embryos using current technologies could have unpredictable effects on future generations." In the United States, the Dickey–Wicker Amendment (1995) prohibits the creation of human embryos for research especially with federal funds. The law in pertinent states

> SEC. 509. (a) None of the funds made available in this Act may be used for
> (1) The creation of a human embryo or embryos for research purposes; or
> (2) research in which a human embryo or embryos are destroyed, discarded, or knowingly subjected to risk of injury or death greater than that allowed for research on fetuses in utero under 45 CFR 46.208(a)(2) and Section 498(b) of the Public *Health Service Act* (42 U.S.C. 289g(b)) (Title 42, Section 289g(b), United States Code).
> (b) For purposes of this section, the term "human embryo or embryos" includes any organism, not protected as a human subject under 45 CFR 46 (the Human Subject Protection regulations) … that is derived by fertilization, parthenogenesis, cloning, or any other means from one or more human gametes (sperm or egg) or human diploid cells (cells that have two sets of chromosomes, such as somatic cells).

The issue of consent is significant and I believe this book has already offered some reflections. It is worth mentioning here that every human being must be able to give consent for any clinical research. But the question herein is, are genes persons? This question is very controversial because it is not possible and I find this argument to have the "Lake Wobegon effect." But on the whole, the gate-keeper theory seems to be a valid natural argument in discussing the limits of GE technologies.

The second premise for the argument in favor of moratorium on human embryonic editing is based on clinical or therapeutic benefits. It has been suggested that the moratorium should be in effect due to "a current lack of compelling medical applications justifying the use of CRISPR Cas9 in embryos." This is a very strong statement in favor of the moratorium. The fact is that gene editing is relatively new and compared to other medical technologies, there are relatively sparse clinical data and research to substantiate its continual applications especially human embryos both *in vivo* and *ex vivo*. This argument does not denounce, however, that the current researches and publications be discarded; it seems to raise a valid existential issue of clinical value. It seems to posit the question of whether current therapeutic interventions are better than the new ones (gene editing). It is a known fact that clinical research on human subjects has been at the epicenter of health care and particularly in ameliorating many diseases. Generally, drugs in Phase III clinical trials are tested on human subjects for "efficacy and potency."[16] Many new invasive therapeutic procedures, too, are tested on human subjects prior to approval. One of the major reasons for clinical research on human subjects is to discover and ascertain a novel way in which a drug, medical device, or therapeutic procedure works in view of treatment of a particular disease or condition. Simply put, clinical trials are *sine qua non* in the advancement of medicine and health care. But such trials can become complicated if there are competing treatments for a particular medical situation. In other words, a situation arises when there is a genuine uncertainty in the mind of the researcher or clinical experts of the "...state of genuine uncertainty regarding the comparative therapeutic merits of the interventions being compared in each arm in the trial."[17] This phenomenon is known as *equipoise*. Equipoise, then, becomes the rational and springboard for randomized clinical trials in order to ascertain the best therapeutic interventions being compared. After all, as the aphorism goes, "it is better to err on the side of caution" than vice versa. The call for moratorium may be significant in light with equipoise. Gene editing must be demonstrated beyond reasonable doubt by practitioners and advocates that it offers better and safe therapeutic benefits than any competing ones. More data will be needed especially in animal models in order to move forward. This leads to another important issue of safety.

So, in April 2015 NIH Director Collins noted that gene editing "offers serious and unquantifiable safety issues."[18] The statement appears a bit amorphous because "unquantifiability" is relative and contingent on available data. Regardless, there is unanimous agreement within the scientific community that there may be potential safety issues in gene editing. Unlike other clinical or therapeutic applications, "germline genome alterations are permanent and heritable, so very, very careful consideration needs to be taken in advance of such applications."[19] The concerns for safety extends to all human applications including specific or clustered genes in humans such as the potential cures for HD, amyloidosis (AL) among others. In addition, formidable international bodies such as UNESCO and WHO have also called for some paucity in gene engineering research so as to give researchers, policymakers, and ethicists the space and some time to reflect on the emerging and potential challenges. UNESCO's International Bioethics Committee (IBC) in fact held an extraordinary meeting recently and issued a document entitled, "*Updating Its Reflection on the Human Genome and Human Rights* ," in which they discussed tersely some of these challenges.[20] The document noted in pertinent part that gene editing could "jeopardize the inherent and therefore equal dignity of all human beings and renew eugenics."[21] Furthermore, the committee was concerned about the phenomenon of "direct-to-consumer" genetic kits easily accessible on the Internet such as 23andMe in which individuals could order their own gene test without ever meeting a physician. The IBC reiterated the cardinal roles of public authorities "...to promote campaigns to inform citizens about the real or unfounded scientific basis of DTC tests and raise appropriate awareness"[22] so as to ensure proper therapeutic use of the technologies. These sentiments have been shared by many other professional bodies such as the AMA, WHO, the WMA, and most fervently by the National Academy of Sciences (NAS), the United States National Academy of Medicine (NAM), the Chinese Academy of Sciences (CAS), and the Royal Society and have convened an international meeting slated in early December 2015. As Paul Nurse, president of the RS, noted: "It is vital that we have a well-informed international debate about the potential benefits and risks, and this summit can hopefully set the tone for that discussion."[23] In fervor of international academic solidarity and prudence, the president of the Chinese Academic of Sciences, Chunli Baid reiterated: "Both Chinese scientists and the government are aware of the pros and cons of human gene editing. CAS scientists have organized a panel discussion and coordinated with related government agencies for regulatory policies on this issue. We would like to work together with international communities for the proper regulation and application of such technology."[24] These statements are significant in a number of ways because the international scientific communities are increasingly recognizing the potential implications of genome engineering, especially in humans, and the growing consensus calling for a halt in order to ensure proper use of the technology as there are genuine safety concerns within the scientific community itself. But in contradistinction to these growing international calls for moratorium, some scholars such as Niakan have submitted that "There are suggestions that the methods could be used to correct genetic defects, to provide disease resistance, or even to introduce novel traits that are not found in humans. It is up to society to decide what is acceptable: science will merely inform what may be possible."[25] The main idea of opposing views is that a ban could impede other numerous applications of the technologies other than therapeutic use and therefore calls for noninterventionist approach to genome engineering. But are these opposing views justified? Why or why not? I will respond to these in the next paragraph.

Noninterventionist or proactionary approach

It has often been suggested that intellectual freedom should not be regulated or impeded in any way with any norm especially, in the scientific and technological arena. Proponents of the *proactionary approach* (which is diametrically opposed to the precautionary approach) argue for some form of noninterventionism where scientists undertake their research at their pace unrestricted because for the most part, science as a genre is never completely understood until after the facts. One of the proponents and advocates of this principle, Max Moore sums it up in these contentious words:

> People's freedom to innovate technologically is highly valuable, even critical, to humanity. This implies a range of responsibilities for those considering whether and how to develop, deploy, or restrict new technologies. Assess risks and opportunities using an objective, open, and comprehensive, yet simple decision process based on science rather than collective emotional reactions. Account for the costs of restrictions and lost opportunities as fully as direct effects. Favor measures that are proportionate to the probability and magnitude of impacts, and that have the highest payoff relative to their costs. Give a high priority to people's freedom to learn, innovate, and advance.[26]

This view is increasingly becoming popular and perhaps relevant in the quagmires of gene modifications. Should gene editing research and applications be allowed instead of restrictions imposed on it? Would a moratorium impede innovations and the potential utilities of the technologies? These and a number of questions have galvanized some scholars and in individuals to vouch for unrestricted biomedical research and applications of gene editing irrespective of the consequences. In a rather tantalizing statement, some research entities in the United Kingdom such as the Welcome Trust, Medical Research Council (MRC), and the Biotechnology and Biological Research Sciences Council (BBSRC) have issued a joint statement postulating the thesis for the noninterventions of gene editing, especially those involving human embryos because these researches have "tremendous value to basic research." The statement continued

> It is important to emphasise that the science is still at a relatively early stage and potential therapeutic applications are not yet here. It is also important to clearly delineate the different ways and contexts in which this technology might be used: clearly distinguishing the use of this technology in a research context compared with its potential application in a clinical setting; as well as distinguishing the use of these technologies using somatic (nonreproductive) or germ (reproductive) cells.[27]

The research groups affirm the general notion that gene engineering holds tremendous benefits to research but the current research is relatively new and has to continue. They postulated the significance in distinguishing gene editing in basic research and clinical applications. They seem to favor unimpeded basic research involving GE technologies especially in nonreproductive cells because "research using genome editing tools holds the potential to significantly progress our understanding of many key processes in biology, health and disease and for this reason we believe that responsibly conducted research of this type, which is scientifically and ethically rigorous and in line with current legal and regulatory frameworks, should be allowed to proceed." In a rather tantalizing statement, George Church has expressed his unalloyed thesis for less restriction in gene modifications involving human embryos. He notes "...even when technology is going very fast, we have tried and tested traditional ways of reining it in. We don't need special bans or a moratorium—we have the Environmental Protection Agency, we have the FDA. We need to think big, but also think carefully."[28] That is to say, there is enough regulatory oversight already, so enacting more laws and moratoriums may probably stall the progress of science. But are these assertions accurate?

Second, it is speculated that any potential ban on GE in humans could impede scientific innovations and the potential therapeutic benefits this may offer may be lost. Alarmist publications with titles such as "Eugenics lurk in the shadow of CRISPR" and "Ethics of embryo editing paper divides scientists" are fueling the debate on the ban without due considerations to the inherent implication on researches involving gene engineering. And as Niakian said recently, "The knowledge we acquire will be very important for understanding how a healthy human embryo develops, and this will inform our understanding of the causes of miscarriage. It is not a slippery slope [towards designer babies] because the UK has very tight regulation in this area."[29] But are scientists trying to design human beings as seemingly postulated in the media especially nonscientific journals or papers? It is important to indicate that currently, three-person IVF or mitochondrial transfer has been approved for clinical applications in most countries such as the United Kingdom and the United States, which involves some form of modifications to the human embryos. In brief, proponents of the proactional approach argue that freedom, a fundamental right of every human being should extend to the pinnacles of academia and obviously, biomedical research in gene editing, in particular, without the possibility of interference or regulatory oversights.

Third, science is self-regulated and there are already many international norms and administrative structures such as IRB's with the preponderance and fiduciary responsibilities of protecting vulnerable populations as well as protecting public safety. Some of the regulations such as the NC and the DOH have emerged to curtail some of the challenges associated with clinical research and offers some general framework for both researchers as well. Since the proliferation of genetic research, the scientific community has been responsive in organizing many forums to deliberate on the potential and actual challenges these scope of knowledge could posit. For instance, the 1970s were markedly fruitful for molecular biologists since the first reliable cloning technology was used and many within the scientific community had raised concerns about safety. Paul Berg, one of the pioneers in rDNA biotechnology helped organize the Asilomar Conference Center, California, in 1975 to discuss the potential biohazards of recombinant biotechnology process such as the use of SV40 (which was considered to cause tumors in mice models). The scientists (140 of them), lawyers as well as some policy makers converged and called for self-regulation of recombinant biotechnology given some of the potential biosafety concerns. The following year, 1976, the US NIH, the Department of Agriculture, the US Environmental Protection Agency (EPA) among others issued official guidelines on the use of rDNA technologies. On January 29, 2000, The Cartagena Protocol on Biosafety was promulgated to regulate the then burgeoning international rDNA or bioengineering industry as well as strengthening existing norms. Also, in view of the heightened debates and concerns about the applications of gene modifications methods, prominent scientists such as Jennifer Doudna, Feng Zhang, and Ron Weiss Carrol Donna participated in a scientific conference in Napa, California, to deliberate on the challenges of gene editing and especially some of the ethical and societal concerns of the technology in general. Another and similar meeting also took place on September 24, 2015, at Cold Spring Harbor with a focus on the topic: *Genome Engineering: The CRISPR/Cas Revolution* with leading scientists such as Doudna participating. It is significant to point out that these meetings have been convened at the behest of scientists themselves to explore the challenges their innovations are positing to society. Furthermore, a corpus of international scientists had converged in Washington DC to deliberate and issue some guidance ostensibly to regulate the gene modification biotechnologies. They called for a moratorium in 2016. However, at the beginning of 2017, the National Academies of Sciences recommended that under certain circumstances, inheritable genes could be edited. Proponents of the proactionary approach may find credence to support their thesis as well as some credulity in the precautionary approach because these initiatives are being taken by scientists themselves. It has been argued that a moratorium could impede scientific advancement; rather, current gene modification technologies should continue unabated because science is capable of internally regulating itself in a bid to protect the greater good of the public. But historical precedents evidenced in the Tuskegee Syphilis Study, Nazi Medical Research on vulnerable people, forced eugenics are concrete reminders of why medical research involving genome editing should be debated with the dexterity and the guidance needed to proceed. That is, science needs to be regulated both from *within* and *without*. It is a fact that GE is relatively new and the call for paucity on human applications are genuinely rooted in the collective responsibility of society and academic and research institutions to assuage the fears of error in genome editing. This view has been supported and expounded by Doudna, one of the coinventors of the CRISP Cas9 systems, at the meeting at Napa. In unequivocal terms, she urged for caution and prudence especially in clinical applications. She indicated "it would be necessary to decide, for each potential application, whether the risks outweigh the possible benefit to a patient. I think this assessment must be made on a case-by-case basis." In brief, rather than a complete ban or excessive control of the gene editing technologies, perhaps a better approach is to allow both the clinical and nonclinical applications on the expediency of each case and contexts devoid of restrictions (in far as no existing regulations are broken and no humans at risk). For example, Layla Richards was recently treated for leukemia in Great Ormond Street Hospital (GOSH) in London. Using TALEN gene biotechnology researchers modified donated T-cells to target Layla's leukemia cells. The gene therapy has been deemed successful but the attending physicians have said "as this was the first time that the treatment had been used, we didn't know if or when it would work and so we were over the moon when it did. Her leukemia was so aggressive that such a response is almost a miracle."[30] Layla's case seems to affirm Doudna and other researchers' views that a complete ban might not be feasible rather each case be treated as appropriate to its context.

In addition, there seem to be competing ethical norms and regulatory frameworks on this matter both within the scientific community and outside of it even so among ethicists. Some ethicists such as John Harris have argued that a moratorium and stringent regulations might have any ethical justifications because "The human genome is not perfect" and "It's ethically imperative to positively support this technology" and therefore does not "...see any justification for a moratorium on research." In other words, gene technologies would enhance the human species if allowed to be used. This view has been accentuated by Neuhauser when he made a comparative observation about those gene modification technologies with IVF: "It was the same with IVF when it first happened. We never really knew if that baby was going to be healthy at 40 or 50 years. But someone had to take the plunge." In their views, the time to make that surge or "plunge" is perhaps now. But these purportedly ethical views have been sharply contrasted by many ethical norms discussed earlier. It is also important to note that the NIH and other regulatory bodies especially in the United States are not actually vouching for a complete ban on every biomedical research involving the use of gene editing technologies; rather the *raison d'etre* for the ban is about gene modification in human embryos because there is no consent from "future generations" for the alterations of their genes in the present.

In brief, three approaches have emerged in the debate on how to proceed with gene modification technologies especially in humans. This has gained both national and international debates. It is the submission of this book that a thorough, open-minded discourse and the willingness to assess each therapeutic applications of the technology will be undoubtedly decisive. A time-restraint moratorium and a conscious effort at progressive discussions of the genetic bioengineering will be in the interest of scientific advancements as well as enhancing better health care. Could a moratorium be another "Asilomar moment" for researchers and other experts to recalibrate data, discrepancies in their researchers, identify potential limits on their innovations; examine some of the ethical trepidations in order to harness the full potentials of bioengineering toward precision medicine and pharmacological developments?

End notes

1 Thomas Kuhn, The Structure of Scientific Revolutions, *The Encyclopedia of Unified Science*, ed. Otto Neurath (University of Chicago Press; Chicago, IL, 1972).

2. A. Ahmed. Precautionary Principle, *the Berkshire Encyclopedia of Sustainability: Vol. 3: The Law and Politics of Sustainability* (Berkshire Publishing Group; Chicago): pp 430–433; Stephen M. Gardine. A core precautionary principle, *Journal of Political Philosophy* 14(1): March 2006, 33–60; Bárbara Osimani. An epistemic analysis of the precautionary principle, *Dilemata* 11: 2013, 149; VosAlles et al. The precautionary principle and the uncertainty paradox, *Journal of Risk Research* 9(4): June 1, 2006, 313.
3. VosAlles et al. The precautionary principle and the uncertainty paradox, *Journal of Risk Research* 9(4): June 1, 2006, 313. See also J. Morris. Defining the precautionary principle, ed. J. Morris, *Rethinking Risk and the Precautionary Principle* (Butterworth-Heinmann; Oxford, 2000): pp 1–21.
4. De Souza, Natalie. Improving gene-editing nucleases, *Nature Methods* 9(6): June 2012, 536.
5. Mali et al. Cas9 as a versatile tool for engineering biology, *Nature Methods* 10(10): October 2013, 957–963.
6. David Benjamin et al. Therapeutic genome editing: Prospects and challenges, *Nature Medicine* 21: February 5, 2015, 121–131.
7. Meeri Kim. Scientists are growing anxious about genome-editing tools, *The Washington Post* May 18, 2015.
8. Ian Sample. UK scientists seek permission to genetically modify human embryos, *The Guardian* September 17, 2015.
9. Rebekah Marcarelli. Genetically modified human embryos could be used for research in the near future, *HNGN* September 21, 2015.
10. www.nih.gov/about-nih/who-we-are/nih-director/statements/statement-nih-funding-research-using-gene-editing-technologies-human-embryoss.
11. David Cyranoski. Scientists sound alarm over DNA editing of human embryos, *Nature* March 12, 2015.
12. Kuhn, pp. 79–80.
13. Ibid.
14. Ibid. Helen Shen First monkeys with customized mutations born, *Nature* January 30: 2014; Helen Shen CRISPR technology leaps from lab to industry (December 3, 2013); Heidi Ledford. Targeted gene editing enters clinic, *Nature* 471: March 1, 2011.
15. Igor P. Pogribny. Alterations in DNA methylation resulting from exposure to chemical carcinogens, *Cancer Letters* 334: June 28, 2013, 39–45.
16. www.fda.gov/clinicaltrials. See also Emmanuel A. Kornyo. Some ethical paradigms in view of equipoise and randomized clinical trials, *Voices in Bioethics* July 7: 2014.
17. Salim Daya. Clinical equipoise, *Evidence-based Obstetrics and Gynecology* 6: 2004, 1. See also, B. Freedman. Equipoise and the ethics of clinical research, *New England Journal of Medicine* 317: 1987, 141–145. See also B. Freedman, C. Weijer, and K.C. Glass. Placebo orthodoxy in clinical research I: Empirical and methodological myths, *Journal of Law, Medicine and Ethics* 24: 1996, 243–251; C. Weijer. Thinking clearly about research risk: Implications of the work of

Benjamin Freedman, *IRB: A Review of Human Subjects Research* 21(6): 1999, 1–5 and C. Fried, *Medical Experimentation: Personal Integrity and Social Research* (American Elsevier; New York, NY, 1974).

18. www.nih.gov/about-nih/who-we-are/nih-director/statements/statement-nih-funding-research-using-gene-editing-technologies-human-embryos.
19. Ibid. Dana Carroll et al. A prudent path forward for genomic engineering and germline gene modification, *Science* April: 2015.
20. Report of the IBC on updating its reflection on the Human http://unesdoc.unesco.org/images/0023/002332/233258e.pdf
21. Ibid.
22. Ibid.
23. International Summit on Human Gene Editing—Home.
24. Ibid.
25. Ian Sample. UK scientists seek permission to genetically modify human embryos, *The Guardian* September 17: 2015.
26. Max Moore. The Proactionary Principle http://strategicphilosophy.blogspot.in/2008/03/proactionary-principle-march-2008.html
27. Genome editing in human cells—initial joint statement.
28. George Church, http://insights.bio/cell-and-gene-therapy-insights/2015/09/21/is-human-embryo-gene-editing-using-crisprcas9-on-the-cards-in-the-uk/.
29. Ian Sample.
30. Ian Sample.

8 Some Perspectives and Conclusion

In contemporary times, molecular genetics, especially the sequencing and study of the genetic underpinnings of the etiology of many diseases, is changing the healthcare landscape. Bioinformatics technologies have improved tremendously, coupled with the ability to process high throughput genetic data. In addition, the costs and time to process whole genomes have also improved. For instance in 2001, the cost per a whole genomic sequencing of eukaryotes was estimated at over $100 million but reduced to about $5000 in 2014![1] It is anticipated that the price of sequencing could further drop to less than $1000 concurrent with a processing time of a few hours. Furthermore, even though the entire human genome has been sequenced, the volume of biodata generated remains magnanimous. While biomedical researchers are perusing and assiduously striving to identify the intricate function and correlation of the genetic constituents of the human genome for the next generation of biologics and precision medicine, it nonetheless, continues to generate many discussions. As indicated in the book, this will be significant in the development of pharmacogenomics and personalized therapy. But such innovations are also fraught with ethical conundrums such as a reevaluation of the question of autonomy, confidentiality, privacy, genetic reductionism, and a concatenation of others.

This book has postulated the thesis for AC in the management of genetic data and biospecimen akin to HIPAA. This is because genetic materials are shared biologic entities; people within specific geophysical and socio-anthropological loci have starkly similar and unique biomarkers and other genotypic and phenotypic identifiers. Given scientific historical precedents of the eugenics movements and furtive research activities, it is important to protect vulnerable populations that might have some genetic dispositions to atypical diseases. In addition, there have been some axiomatic shifts from paternalism in medicine to patients having prerogatives (autonomy) in deciding their medical needs. Consequently, patients and their proxies may be encouraged to inform their familial relations of potential genetic risks. Also, this book argues for the absolute de-identification of human subject genomic data in pharmacogenomics research especially those that might be published or made available in some form of public domains in order to insulate them from potential genetic stigmatization.

In addition, in an information-driven era, *en mass* national and international electronic data (ED) could be an integral part of clinical practices. ED could coalesce biomedical records into a single technology platform that is accessible to healthcare practitioners and possibly available for pharmacogenomics study. Such electronic databases will have, among other things, whole genomic sequences (WGS) of every consented patient both at birth and when they become adults. I believe this will provide an efficient diagnostic and prognostic tool especially in times of dire medical emergencies and other exigencies. It is important to support already existing disease-specific groups/foundations such as the RARE Foundation Alliance (which seems to be an amalgamation of all known rare genetic disease groups) in order to develop better pharmacogenomics data sharing in a bid to offer personalized care. Also, there are many genetic databases (albeit fragmented) for research purposes such as Gene Disease Data, SNPedia, National Microbial Pathogen Data Resource, and Bioinformatic Harvester. These genetic-based databases will continue to be very relevant in biomedical research. It is however, important to ensure that there are proper regulations of these to ensure that the data or the bioinformatics that they provide are accurate and have been ethically generated.

Finally, due to the sociocultural sensitive nature genomic studies are generating, it will be important to incorporate socio-anthropological training and ongoing education for medical and allied biomedical professionals. Genetics is just part of the whole person. There are obvious cultural and environmental indices and long historic factors that are critical in overall health care. While such trainings are currently being touted in the training of new generation of physicians and healthcare professionals, more has to be done. Could half the period for physician-residency program be spent on medical anthropological field training?

Indeed, a holistic personalized care should be within the context of a strong regulatory, legal and socio-anthropological, and ethical framework of respect for the autonomy of patients, protection of vulnerable populations among others. A good precision medicine will begin with a good anthropological framework that truly integrates all aspects of a person's experience. As Hippocrates once noted, "It's far more important to know what person the disease has than what disease the person has"![2] I believe this new impetus toward the incorporation of the advancements in pharmacogenomics and personalized therapy into general health care will be consistent with the core mandate of the *Hippocratic* tradition.

End notes

1. DNA Sequencing Costs—National Human Genome.... https://www.genome.gov/sequencingcosts/.
2. www.iep.edu/hippocra.

Index

A

B

C

D

H

I

J

L

M

POCKET GUIDES TO
BIOMEDICAL SCIENCES

Series Editor
Dongyou Liu

A Guide to AIDS
Omar Bagasra and Donald Gene Pace

Tumors & Cancers: Brain – Central Nervous System
Dongyou Liu

Tumors & Cancers: Head – Neck – Heart – Lung – Gut
Dongyou Liu

A Guide to Bioethics
Emmanuel A. Kornyo

Tumors and Cancers: Skin – Soft Tissue – Bone – Urogenitals
Dongyou Liu